THE PUZZLE OF
LEFT-HANDEDNESS

The Puzzle of
Left-handedness
Rik Smits

REAKTION BOOKS

Published by
Reaktion Books Ltd
33 Great Sutton Street
London EC1V 0DX, UK

www.reaktionbooks.co.uk

Het raadsel van linkshandigheid was first published by Nieuw Amsterdam,
Netherlands in 2010 as a revised and expanded edition of an earlier edition
entitled *De linkshandige picador*, published in 1993 in the Netherlands by
Nijgh en van Ditmar.

Copyright © Rik Smits 2010

First published in English 2011

English-language translation © Reaktion Books Ltd 2011

English-language translation by Liz Waters

This publication has been made possible with financial support from the
Dutch Foundation for Literature

Printed and bound in Great Britain
by the MPG Books Group

British Library Cataloguing in Publication Data
Smits, Rik.
The puzzle of left-handedness.
1. Left- and right-handedness.
I. Title
152.3'35-DC22

ISBN 978 1 86189 873 9

Contents

1

Malice and Misunderstanding

'My husband's left-handed too.' The young woman sounded worried. 'Is he going to die young?' I couldn't help laughing; here we go again. The story that the average left-hander dies an untimely death has been around for decades. It's a dark fairy tale that's proven tenacious for two reasons. Every once in a while a scientist comes along who's happy to breathe fresh life into it simply for the sake of causing a sensation, but more important, perhaps, is the eagerness with which the vast majority of right-handed people embrace the story every time it crops up. What could be nicer than a good safe shudder? Whether or not it's true isn't really the point. It's mainly a matter of guilt-free gloating.

The ease with which such stories circulate on the global rumour mill suggests that people don't really believe them. The version most commonly heard is that left-handers die an average of nine years sooner, and that by about the age of forty most have disappeared. Were that truly the case, then left-handedness would be an extremely grave affliction, a scourge claiming a huge number of young lives – the kind of illness people talk about in veiled terms, preferably in acronyms.

'Have you heard? Caroline's got LH!'

'What? Her too?'

'Yes. You didn't hear this from me, but ...'

'Gee ... That family sure has it hard.'

This is not the sort of thing people say.

It all started in 1992, when a Canadian professor of psychology by the name of Stanley Coren published a book in which, based on his own research, he drew attention to the horrifying fact that left-handed people have lifespans nine years shorter than average. He called it *The Left-Hander Syndrome*. There's no denying that Coren had a feeling for drama; he made rather a name for himself with his book and it earned him a tidy fortune. Then people simply got on with their lives. No politicians asked questions in the House; no specialists appeared in the media to inform the public. We didn't even see any worried left-handers forming action groups, or fearful parents-to-be demanding early prenatal testing for this fatal disorder. Nobody pressed for further research. Nowhere in the world was there even a well-meaning government information campaign. Speculation about left-handed mortality was and remains public entertainment – which is not something real illnesses are.

Coren's book fitted a pattern that often emerges when left-handedness makes news: something grabs our attention, but it's never more than a flash in the pan and left-handers themselves usually decline to comment. That's another thing right-handed people always seem to overlook. In 1998 Australian biologist and journalist Geoff Burchfield asked the psychologist Michael Corballis in an interview: 'Why do you think the subject of handedness triggers such strong responses in people, especially lefties?' Corballis began in all seriousness to articulate a detailed answer, but he soon got bogged down in meaningless platitudes. Not that Corballis is stupid, far from it, but the question was as natural as it was misconceived. The unconventional preference exhibited by left-handed people, whom Burchfield rather dismissively refers to as 'lefties', is in reality something that astonishes and preoccupies right-handers. They regard the left-handers they come across from time to time as peculiar – and indeed intriguing and faintly creepy – whereas for left-handed people it's the most normal thing in the world. They don't see themselves and their like as in any way odd, and they are entirely used to living in a right-handed world. From an early age they have grown accustomed to the strange fascination their aberrant hand preference engenders in right-handed people.

This is a pity, really, because although right-handers get thoroughly het up about the problems they perceive to exist while left-handers merely shrug, all sorts of real puzzles remain. Along with language and cooking, the fact that the vast majority of human beings have a distinct preference for the right hand is one of relatively few characteristics that are unique to our species. Preference for one hand or paw over the other is not unusual in itself – in fact it can be found in a wide range of animals, even in mice – but the uneven distribution of left- and right-handedness is a different matter. Only one in ten people regard themselves as left-handed, whereas among animals it seems that a roughly equal number favour the left paw as favour the right.

No one can yet say with any certainty just how it comes about that one person is right-handed and another left-handed. Equally puzzling are the reasons why left-handedness arose in the first place. It must have happened a very long time ago, because another puzzle is why down through the centuries, all over the world, there has always been a stable left-handed minority of around ten per cent. That's not all. Hand preference reflects a deeper, all-pervasive quality of human beings, namely the fact that each cerebral hemisphere has its own distinct functions, yet in most left-handed people the layout of the brain seems to differ little if at all from that of right-handed people.

As if all this wasn't confusing enough, the concepts 'left' and 'right' are problematic in themselves. We find it hard to learn which is which, and even in adulthood we can quite easily mistake one for the other. Driving instructors are all too well aware of the possibility that a nervous pupil will yank the steering wheel in the wrong direction. Nevertheless, there are innumerable ways in which the distinction between left and right determines how we order, experience and comprehend the world. Photographs, drawings and paintings comply with implicit laws of orientation, as do films and comic strips.

Despite how successful some left-handers are, and how inconspicuous most remain, throughout history left-handedness has been associated with clumsiness and with unpleasant traits such as untrustworthiness and insincerity. The Latin word for left, *sinister*, has all kinds of dark, dismal and ominous connotations that have come into every one of the many languages related to Latin. Science has thrown its weight into the ring at regular intervals. At the end of the nineteenth century infamous skull-measurer Cesare Lombroso had no hesitation in saying that left-handedness was a sign of a criminal personality, and in the mid-twentieth century prominent American psychoanalyst Abram Blau announced that left-handedness was tantamount to 'infantile negativism', the equivalent of a refusal to eat everything on your plate. Neither of these eminent gentlemen had the slightest evidence for their harsh judgements, but they weren't going to let that spoil their fun.

The actions and attitudes of Coren and his ilk demonstrate that as far as rushing to conclusions goes, little has changed. Left-handedness was and is associated with maladies of all kinds, including mental retardation, alcoholism, asthma, hay fever, homosexuality, cancer, diabetes, insomnia, suicidal urges and criminality. In most cases there's a complete absence of solid evidence for any such association, although it is true that groups of people with minor ailments, disabilities or stains on their

The notorious Italian skull-measurer Cesare Lombroso, who eventually also came to believe in ghosts.

character generally include more left-handers than average, whether they be hay-fever sufferers, breast cancer patients or prison inmates. But higher than average percentages of left-handers can also be found in art colleges and architecture schools, and among the highly gifted. Then again, it turns out that within randomly composed groups of people, no differentiation based on character traits or failings has ever been found to coincide with a division into left-handed and right-handed.

The puzzles and paradoxes continue to stack up. There are plenty of indications that left-handed people are entirely normal, except in their hand preference, yet the fuss that's made even today about writing with the left hand implies otherwise. It's the subject of endless philosophizing and theorizing and often described as contrary to nature. Much is made of the 'serious problems' faced by left-handed six-year-olds who are called upon to 'imitate right-handed writing using their left hands', as Dutch handwriting guru A. van Engen puts it. Oddly, despite all this concern, teacher training colleges fail to pay serious attention to left-handedness even today. Left-handed six-year-olds generally have to find out for themselves how to master handwriting – the most difficult skill taught in primary school – based on a model that shows them the reverse of what they are trying to do. Most manage it so successfully, despite all the problems they're presumed to encounter, that in no time at all they can write as well and as quickly as their right-handed classmates.

Of all the many paradoxes, this one is right under our noses, since what could be more easily observable than a child writing at a classroom desk? Unless you believe in miracles, only one conclusion is possible: the image we have of left-handedness is based only to a small extent on the actual behaviour of left-handed people. Our attitudes are derived from other magical, mythical, traditional assumptions about left and right, which must be deeply rooted in our habits of thought. To unravel the puzzle of left-handedness we will have to start by examining these culturally determined assumptions. To be precise, we must return to the eve of the twentieth century in the Carrer dels Escudellers Blancs in Barcelona, to the studio of a young hothead called Pablo Picasso.

2

The Left-handed Picador

In 1899 Pablo Picasso, at eighteen already a reasonably successful up-and-coming artist, tried his hand at copper-plate engraving for the first time. He created a standing portrait of a picador, the man at a bullfight who rides around on horseback and goads the bull with a lance. The result was a disappointment in every respect, but most of all because it seemed the lance had unintentionally been placed in the picador's left hand. Of course Picasso had engraved an appropriately right-handed picador, but his inexperience was such that he hadn't taken account of the fact that an etching is always a mirror image of the original. He cleverly made a virtue of necessity by inscribing in wild lettering above the final result *El Zurdo*, the left-handed man. His honour was spared, but it was another five years before Picasso ventured to make another copper-plate engraving.

Picasso's sense of disappointment shows how deeply ingrained is the distinction between left and right; how attached we are to getting things the right way around; how eager we all are, unconventional artists included, to conform to the norm: right-handedness.

The importance of choosing the right side in the most literal sense was demonstrated by the threats aimed at the world community, over the heads of Congress, by us president George W. Bush on 20 September 2001, just a week after the attack of 11 September 2001 that destroyed the World Trade Center in New York, with almost 3,000 deaths as a direct consequence. 'Every nation,' he said, 'in every region, now has a decision to make. Either you are with us, or you are with the terrorists.' On 6 November that same year, before preparations began for the invasion of Iraq – which he unleashed in March 2003 in alliance with what became known as the 'coalition of the willing' – Bush once again made clear just how simple a situation this was: 'Over

Pablo Picasso, *El Zurdo*, 1899.

time it's going to be important for nations to know they will be held accountable for inactivity. You're either with us or against us in the fight against terror.'

Bush was the target of much justified criticism for his black-and-white vision of world politics. Anyone who kept a cool head for a moment and gave the matter some thought understood that there were all kinds of reasons why a country might decline to join Bush's punishment expedition without necessarily harbouring any sympathy for the enemies of America, but the president, in his intellectual simplicity, provided a clear example of how people generally tend to think: in black and white. A whole arsenal of sayings underlines the point: the best of both worlds, it's a two-way street, boom or bust, stand or fall, do or die. People are all too keen to make such distinctions as clear-cut and absolute as possible. Black-and-white clarity gives us more confidence, a greater feeling of having a grip on reality, than the grey tones of accuracy and nuance.

The US president was hardly the first leader to express himself in this way. For all his lack of charisma and rhetorical talent, Bush's oversimplified statements placed him in a long, motley line of demagogues that goes back at least to Alcibiades of Athens in around 450 BC. He understood perfectly well, as did more recent historical figures such as French revolutionaries Danton and Marat, Lenin, Hitler and Mussolini, and indeed the entire gamut of present-day populist strongmen, that playing the crowd is all about polarization, the distilling of complicated issues into a simple antithesis. History is full of variations on the theme of 'we're in the right so they're in the wrong': we proletarians are honest, poor and oppressed, so every non-proletarian is an lackey of deceitful, filthy rich and oppressive capitalism; we Westerners love freedom, therefore the Communists tried to crush us. Similarly, most religions, especially the great monotheisms, keep their flocks united by telling them they've been chosen by the one true God and everyone else is doomed, or at least inferior. This is as true of Judaism as it is of Islam. Even Christianity, which has made a doctrine out of charity and mercy, has its Day of Judgment when the sheep and goats will be separated for all eternity.

Just how naturally polarization comes to us was illustrated by the witch-hunt against people alleged to have leftist sympathies in the United States of the 1950s. The episode is known as McCarthyism, but Senator Joseph McCarthy was in reality no more than a willing camp follower who managed to profit from a fear of Communism that had been growing steadily since the Second World War. Stalin's Soviet Union had demonstrated its military might during that war and, mindful of the revolutionary Bolshevik rhetoric of the 1920s and '30s, America was terrified of a Communist coup, even invasion. People started to see spies everywhere and in 1947 the US government initiated so-called loyalty reviews. Congress, not wanting to be left behind, set up its own commission to track down disloyal elements. Its most zealous member was a young, ambitious politician called Richard M. Nixon who many years later, as US president, would be brought down by his mistrust of others. It was called the House Committee on Un-American Activities. Note that Congress did not choose to talk of 'anti-American' or 'pro-Communist' activities. This is the most primitive and at the same time the ultimate form of black-and-white thinking: put 'un-' in front of your ideal and you know what it is you need to combat.

All this suggests that human beings are simply not cut out to handle nuances, since our way of thinking is based on dualism and dichotomy. Perhaps our approach has to do with the fact that there are two sexes, or maybe it arises from the distinction between the self and the rest of the world. It may have its origins in something else entirely, but the fact is that we start out by attempting to reduce any complex matter to a distinction that lies within a single dimension, imaginable as a line. We then pick a criterion and use it to chop that line in two. Every phenomenon and every property of nature is dealt with in this way: vertical length is divided into tall and short; bulk into thick and thin; time into early and late. It's no different with man-made concepts that don't exist in the natural world. Things are good or bad, beautiful or ugly, pleasant or unpleasant, true or false.

Triality is unknown. There's no obvious concept of the same order to set beside true and false, or high and low. Even dealing with two dimensions at the same time, such as breadth and depth, is too much for our simple brains. What do we mean by a balcony that's a metre and a half wide? It could be a robust structure projecting a generous metre and a half out from the wall, or a measly strip of decking attached to just a metre and a half of the facade. We learn to cope with ambiguities like this, but we always have to give them a moment's conscious thought and we regularly make mistakes, which estate agents are happy to repeat, or indeed exploit.

Of course we wouldn't get far if we were capable only of thinking in crude dichotomies, but we can refine our world view considerably by dividing one of two parts into two again. For example, once we've made a distinction between edible and non-edible things, we can split the first category into 'tasty' and 'foul'. Division is a recursive process; you can go on bifurcating the result time and again. Fortunately this means that a primitive splitting method is all we need to build up an extremely fine-grained picture of the world.

Not all such dichotomies are of the same kind. Most split the aspect of the world to which we apply them into two parts of random size. For most people, the category 'edible things' will contain far more foods they like than foods they dislike. This type of division does not tell us anything about the content of the two parts: one person regards braised pig's stomach as a delicacy and shudders at the thought of a hamburger; another has precisely the opposite reaction. Any vegetarian worth his salt will turn up his nose at either. So people who apply this kind of distinction decide for themselves exactly what belongs on one side or the other and therefore how large each category will be.

As well as this sort of arbitrary dichotomy there are symmetrical divisions that always by definition produce two parts of roughly equal size, and where demands can be made of the characteristics of each half independently of the person making the split. Examples include front-back and top-bottom. The top part of a person runs roughly from the crown to the navel, never from the crown to the knees. The bottom part of a dog contains everything from its toes to an imaginary line running more or less from its breast bone to its anus. The tail, for example, does not typically belong to the lower half of a dog even if it hangs down. Something similar applies to front and back. Here too we divide a person, animal or thing into two halves of roughly equal size.

With most inherently fifty-fifty divisions, the content of the two halves will differ in a clear and important sense. A ball does not have a top or a back, simply because there is no demonstrable difference between one part of a ball and another. If we do talk about the back of a ball, we don't mean any specific part but simply the bit that happens to be invisible from our point of view at that particular moment. Trees have a clear and inherent top and bottom; even if we turn a tree upside down, the roots, or the trunk, still belong to the lower part. But a tree, like a ball, does not have a true back and front of its own.

The pair left and right are a special case of this kind of division. Most living creatures visible to the naked eye exhibit a clear distinction between their upper and lower halves, both in the functions concentrated there and in appearance. Most animals, and almost all vertebrates, have a front and back that are clearly differentiated. Those two dimensions are easy to recognize according to explicit criteria. Gravity defines the vertical dimension, while front and back correlate with 'towards us' and 'away from us'. The side we're looking at as something moves towards us is called its front and as it moves away from us we see its back. In the case of immovable objects, the front is the side we normally see when we move towards them. So the front of a house is on the street side where, as a visitor, you ring the bell. The front of a dog or a ship is what we see as they approach us; the back of a drill is the part we're facing as we push the thing into a plank or a wall.

No such criteria exist when it comes to left and right, which explains why we have difficulty with them as a conceptual duo. Animals and plants are generally symmetrical as far as their left-right axis goes, although this is not at all the same thing as the dull uniformity without qualities that's a feature of the surface of a ball. As a rule the right and left halves are outwardly almost complete opposites, yet they differ only by a hair. Our one-handedness proves that uniformity is not the same thing as equivalence. Since human beings are born splitters, it's no wonder we're intrigued by this imbalance, nor that left, right and symmetry have become crucially important in such characteristically human products as works of art, handwriting and the symbolism of our world view.

3
Opposites and Contradictions

Deep in the mists of time, more than 3,000 years ago, Greece must have been inhabited by a farming folk that worshipped earth gods, first among them the earth itself, the fruitful mother in whom all life originated. It's therefore commonly assumed they were a matriarchal people, with a society in which women, bound to Mother Earth, were in charge. However that may be, one ill-fated day Indo-European nomads invaded Greece. These sturdy warriors had little difficulty overrunning the earlier inhabitants, and they entertained very different beliefs. In their experience the earth was relatively unimportant. What mattered to them were the open horizon, travel, hunting and warfare. Their society, far from assigning women a leading role, allowed men to make all the decisions. The world of the gods is always a reflection of the human world, and the Indo-European deities were no exception. They were mostly men, and they personified powers such as the sun, light and wind. They resided not in the warm darkness of the earth but high in the sky.

The invaders settled permanently in their newly conquered lands and slowly merged with what remained of the original population. After a while, the only traces left of the drama of invasion were stories, and no one any longer had any clue how much of what they told each other was true and how much invented. So as time went on history turned into mythology. People became heroes, and heroes gradually assumed divine proportions.

Something similar happened with the contrasting worlds of the old and new gods. Religions are tenacious, so instead of disappearing, all kinds of elements from the ancient earth-god faiths were merged into the new panoply of Indo-European gods. In classical mythology this process left its mark in the strange and sometimes contradictory family relationships between the many gods and demigods.

The result was a bipolar divine world, dominated by the Olympian gods of heaven headed by their father Zeus but including other important and even more ancient divinities such as the earth-shaker Poseidon, Demeter the goddess of fertility – whose name literally means Mother Earth – and Hades, ruler of the subterranean kingdom of dead souls. Various other ancient cults, such as the worship of the moon goddess Cybele, failed to gain such a prominent place for themselves within the 'official' religion. Gradually, surviving at the margins, they acquired the character of secret societies, which were naturally seen as untrustworthy. They had to be rooted out, even though as time went on no one any longer knew why. An important consequence of this was that darkness, femininity, the earth and fertility became closely associated with intangible mystery, menace, wickedness and magic.

The Indo-Europeans, who came from somewhere in the Near East, didn't all end up in Greece but spread out across Europe and Western Asia, migrating as far as the Indian subcontinent. Everywhere they settled they imposed their norms and values, and again and again these were merged with the remnants of the cultures they had conquered. Right across that vast region, mythologies and religions grew up that in essence had a great deal in common. Whether known by his Greek name of Zeus or, as in Sanskrit, Dyaus Pitar – a name we encounter again in Latin as Jupiter – or, as the Germanic tribes called him, Tiu, the god of gods is always a man. He is the father who sits in majesty high in the sky, associated with the sun, thunder and lightning, and other phenomena of the heavens. Opposing him are the subterranean powers of darkness. They are usually rather suspect and they always take second place, but that certainly doesn't make them insignificant.

The Christian evangelists and missionaries who arrived to convert Europe centuries later had a good deal of fun with all this. They too brought with them a God the father in heaven, and it's surely no coincidence that he too had a tendency to throw bolts of lightning. The foundations had already been laid, in the form of self-evident symbolism in which the concepts of man, master, good, light and heaven belonged together as they do in Christianity. It was easy for the concept of the Devil to develop out of the opposite pole, the earthly darkness, and that symbolism survives to this day in Western cultures in all kinds of ways – in baby clothes, for example, with boys in blue, the colour of the firmament, and girls in pink, associated with blood and the earth.

When the first philosophers, the scientists of this misty antiquity, attempted to understand the phenomena they saw around them, they had no tradition on which to fall back. Everything had to be invented from scratch, and there were few means available other than the already existing system of religious symbols and the philosophers' own dualistic intellect, with its ability to split and to polarize. This produced a set of opposites, and with it a set of connections, that seemed to shed light on the way the world was composed.

One of the most prominent of those early scientists was Pythagoras, who founded a philosophical institute in about 530 BC in Croton, a Greek colony on the eastern coast of the heel of Italy. Today Croton is a remote, rather uninspiring provincial town, but in those days it was a hypermodern city, teeming with creative ingenuity. It was so modern and wealthy, in fact, that it hired professional sprinters and wrestlers from far and wide to enable it to triumph repeatedly at the Olympic Games. Sport was so important that it even led to an all-out war between Croton and its rival Sybaris. Meanwhile Pythagoras and his pupils came up with a number of principles of mathematics and what we would now call music theory. As Pythagoras saw it, everything in the world ultimately turned on numbers and on numerical relationships between whole numbers. The length of a lyre string corresponded to its pitch, and a pleasing relationship existed between the lengths of strings of equal thickness and the harmonious combination of tones they produced. Building on that idea, Pythagoras was able to equate highly diverse matters, in essence, with numerical relationships. The number five, for example, represented marriage, the fusion of the smallest even number with the smallest uneven number larger than one: marriage paired three with two, man with woman, uneven with even.

Croton had its own ideas about all this and they were far from positive. Eventually Pythagoras was forced to flee the city with his followers and for many years after his death the Pythagoreans were actively persecuted. So we cannot credit those ancient sports fanatics of the Italian peninsula for ensuring that some of his work would survive. That was left to others, among them an even greater Greek philosopher, Aristotle, who in his *Metaphysics* adopted a Table of Opposites compiled by Pythagoras. Some of the contrasts it lists are as follows:

even	odd
female	male
darkness	light
evil	good
cold	warm
crooked	straight
left	right

This clearly implies male dominance. Pythagoras regards a man as self-evidently associated with goodness. His counterpart, a woman, is therefore naturally saddled with the opposite of goodness, and this creates a definitive connection between the female and wickedness. In Indo-European cultures, light, the sun and heaven are closely associated with the dominant, male divinities, while darkness and the earth traditionally belong on the female side. Its primeval connection with the female cycle and its marking of important dates in the agricultural calendar ensure that the moon, which shines at night, fits perfectly into this scheme of things.

Now it becomes clear why cold and female are on the same side. Light and the sun, and therefore warmth, are associated with maleness; therefore cold must inevitably belong on the same list as female. It's slightly harder to understand the placing of crooked and straight, but one possible explanation is that to the naked eye there are hardly any straight lines in nature. Straight things are typically man-made. A great deal of effort was required to make an object neat and straight, and if anything went wrong with a piece of work then it turned out crooked. This almost inevitably meant that if something was straight it must be good, since otherwise no one would take so much trouble over it. So straight belonged in the same group as good and was therefore associated with men, while crooked ended up in the same category as women.

Even in this very early symbolic system, the right is on the side of good. For a long time this was thought to have something to do with sun worship. Many ancient peoples orientated themselves towards the east, where the sun rises. In the Arab world this holds true even today. If the east is at the top of a map, then the south, where the sun ensures warmth and light, is to the right. It has been argued that this made the south the good direction, associated with warmth, light, divine assistance and so forth. Polarization did the rest.

Yet this explanation cannot be correct. After all, in the southern hemisphere the sun, although still travelling from east to west, passes to the north rather than the south, so left ought to be regarded as the good side, which is certainly not the case. As far as left and right are concerned, beliefs are no different in the southern hemisphere.

The fact that all over the world people divide things the same way – left for evil and female, right for good and male – suggests that there must be another reason: the dominance of both right-handedness and the male. Right-handers form a large majority of all known peoples. This is enough in itself to explain why the right is more likely than the left to be connected with goodness, since it places most people on the side of good. Furthermore, almost all peoples are patriarchal, so if right is associated with the good then it must also be associated with the male, and with the gods, who are good, or in the case of the Jews with that one nameless god who will tolerate no others. The name by which we still know his great adversary, Satan, is a corruption of the Talmudic Samael, a name derived from the word *se'mol*, meaning left.*

It is ironic that the left became automatically associated with the female, since left-handedness occurs slightly more often in men than in women. It could to a minuscule degree be called a male characteristic. No one seems to have noticed this. Clearly we don't much care whether a view of the world that arises from a symbolic system fits with our actual experience. What matters is the illusion that we comprehend and therefore rule the world, rather than the idea that we have a convincing description of it. We aren't even bothered by the most absurd contradictions. Pythagoras' table, for instance, links women with cold and darkness, the typical characteristics of death, even though the female continues as ever to symbolize fertility and the source of new life. Symbolic systems create order out of the chaos of the world without necessarily entailing anything beyond themselves.

Yet down through the centuries these symbolic systems have unquestionably influenced the way in which we look at the world, and they continue to do so. They form the basis for deeply rooted traditional norms. Women have encountered great difficulties as a result, but so, to a lesser extent, have the left-handed. Some cultures have a real taboo

---

* The Latin word *sinister*, by contrast, gained its dismal connotation only later. The word is derived from *sinus*, a fold on the left side of the Roman toga that served as a pocket. Sinister originally meant simply 'on the pocket side'.

against the left, especially the left hand. Although in large swathes of Europe a person who eats with his left hand can expect nothing worse than a few strange looks, in other cultures, including those of the Islamic world, such behaviour is utterly unacceptable.

4
Taboos, Sex and Handicrafts

In Arab countries the left and right hands have different functions. There too the great majority of people are right-handed, so traditionally the right hand is called upon to do the most important jobs, such as eating, writing and greeting others. The left hand exists to do the opposite, including the dirty work, like the cleaning of the anus. In a culture where people generally eat with their hands, such a division is in fact quite rational, all the more so in the warm climates where Islam has its origins and its greatest distribution. But people will be people, and nothing invites violation so much as a ban based on rational considerations. A taboo works far better, a prohibition grounded in indefinable fear. A taboo it therefore became. The left hand is the unclean hand.

Some claim that within the Islamic world the unclean left hand is intended for the game of love. It's impossible to find any reliable confirmation of this, and anyhow it seems an improbable story. For a start, rules for hand use are a matter of etiquette – you need to learn which tasks you're forbidden to perform with the hand you favour – but it holds true in both European and Islamic countries that the more traditional a belief system is, the more people act as if sex doesn't exist. In the worst case, which is not to say there's anything rare about such things, you're not allowed even to look at girls, so you're hardly likely to be told by someone else which hand to put up your wife or girlfriend's skirt. The same applies the other way around. No decent girl touches a boy with either her right hand or her left. So one taboo obstructs the spread of another.

Of course most people do engage in amorous little games, but they're usually left to discover the rules for themselves, on the sly. Etiquette has far less influence here than the natural limitations of our bodies. A great deal depends on the position of the partners in relation to one another. For a start you have to be able to reach, which is not always possible with the left hand.

Anyhow, I've asked around. No consistent connection emerges between the favoured lovemaking hand, whether for intercourse or for masturbation, and the hand used for other activities, such as writing. Those who wrote left-handedly sometimes made love with their left hands, but by no means always, and vice versa. Most people did say that when masturbating they had a clear and consistent preference for one hand or the other, just as they did when writing. It simply didn't work with the other hand. If that preference is so strong and so independent of other factors, then it's difficult to believe that people, whatever their culture, would override it, especially in the absence of any explicit instructions on the matter.

An apparent exception are the Kaguru, a people of Tanzania. It would seem their youngsters can discuss sex fairly freely between themselves. Young men sometimes boast about how it's best if a man is able to arrange things in bed such that he has his left hand free and the woman her right. Then he can grope at her to his heart's content without using his clean right hand. It sounds tremendously exotic, but such a story raises the question of whether the real point is not so much to use the already dirty left hand to touch the conquered lady as to stage a power display. After all, it leaves her with no choice but to touch him with her clean right hand, whether or not that's something she's happy to do, so there's an implicit humiliation here. The suspicion that this has more to do with macho behaviour than with strict beliefs about cleanliness and uncleanliness is reinforced by the fact that those same boys are happy to admit that the positional preference is really only a vague ideal, which they don't make much effort to put into practice.

Taboos and etiquette are not accorded the same weight at all times. In general we can say that the more formal a situation is, the more strictly the rules are enforced. This is logical, since the more official and staid the occasion, the less well the participants are likely to know each other and the harder it will therefore be for them to set straight any misunderstandings. Clarity and predictability are of immense importance in tricky situations, so we stick to strict protocols, with endless rituals and symbolism. Anyone who takes liberties with the rules will bring down the whole house of cards. This explains why many public figures are quite relaxed about picking their noses at home, or letting out resounding farts, or scratching their heads vigorously, but would never do such things while performing their official functions.

What goes for nose-picking also goes for the taboo against the left side, as the British government discovered to its embarrassment during the Second World War. When Churchill and Roosevelt visited Saudi Arabia for talks with King Ibn Saud, there naturally had to be an exchange of gifts. Churchill promised his host a specially armoured Rolls-Royce, so that at last he could move around safely using a modern mode of transport. King Saud was suitably delighted, if for a quite different reason: such a vehicle would come in very handy for hunting. But when the car was delivered, Ibn Saud realized he'd never be able to use it. As in all English cars, the steering wheel was on the right, so while out hunting the king would be sitting to the left of his chauffeur. For a king that was unthinkable. As in the West, the right side is the place of honour. A disappointed Saud gave the car to his brother Abdullah, who was presumably less bothered about such matters.

Ibn Saud's Rolls-Royce represents a diplomatic blunder of the first order, one that could easily have been avoided if the British foreign office had stopped to think, but sometimes an insult lurks where you'd least expect it. In 1762–3 an expedition dispatched by King Frederick v of Denmark, manned by four scholars and a painter-engraver, travelled through the south-west of the Arabian peninsula, the region known in Arabic as Yemen (literally 'the land of the south') but in Europe called Arabia Felix, or Happy Arabia. The aim was to map the region, make contact with the people who lived there and collect all available facts about the place.

The expedition ended in disaster. Geographer Carsten Niebuhr was the only member of the group to make it back alive, after endless wanderings, and he told the whole story in *Description of Arabia*, published in 1776. In his book he was fairly positive about the Arabs. Anyone who showed proper respect for them could generally depend on being treated decently in return, he wrote. So later expeditions were considerably surprised by the dour, unsympathetic treatment they received. The cause turned out to lie in the maps Niebuhr had made. As was usual in Europe, they were orientated towards the north, so Yemen was on the left side of Arabia. The Yemenis, it transpired, regarded this as an insult. Any right-minded Arab takes his bearings from the east, which puts Yemen, at the south-westerly tip of the peninsula, on the auspicious, right side of the map. This was precisely the reason why in the Latin world the country was called Arabia Felix. On Niebuhr's maps, used by later travellers, it instead lay on the accursed left side.

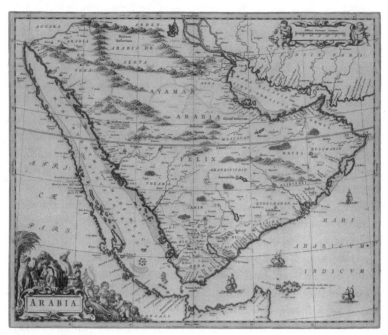

A map of the Arabian peninsula made in about 1800, on which everything currently called Saudi Arabia, plus Oman and the United Arab Emirates on the south-eastern side, is labelled 'Ayaman' or 'Arabia Felix'. That name should really be applied only to the bottom left corner of the peninsula, present-day Yemen.

There's a direct counterpart to the name Yemen, incidentally. The Arab word for Syria is Sam, a word related to *simâl*, which means 'north' and 'left'. Sam is also connected to the verb *sa'ama*, which originally meant both 'to bring bad luck' and 'to turn left' but over time acquired a third meaning: to go to Syria. The connection with bad luck is reflected in many sayings and expressions about the unpleasant effects of the desert winds from the north.

Arab culture is certainly not alone in having a strict taboo against the left side and left-handedness. Japan, hardly a paragon of social flexibility and tolerance, is perhaps the worst case of all. Left-handedness used to be completely unacceptable there and in many ways it still is. In earlier times, left-handed women are said to have concealed their 'abnormality' from their husbands, because left-handedness could be grounds for rejection. Fairly recent research on Japanese schoolchildren suggests that a mere 2 per cent write with their left hands, even though almost everywhere else in the world the number of left-handers is roughly 10 per cent. Japanese researchers have claimed that the low figures are a result of the special qualities of the characters used in the written language, insisting it can be produced only with the right hand. In reality writing Japanese with the left hand isn't particularly difficult. The cause is more likely to lie in the repressive Japanese school system, which simply refuses to tolerate left-handedness. This conjecture is bolstered by the fact that in Europe and the United States far fewer children wrote with their left hands in the past than now, sometimes as few as 2 per cent, as in Japan. When teachers gradually ceased objecting to left-handedness in the classroom, the percentage steadily rose, eventually reaching that magical figure of around 10 per cent.

A similar hostility to the left side and the left hand can be found in many parts of Africa. Often this has to do with the influence of Islam, but not always. Tribes of the lower Niger have a rule that a woman must use her right hand rather than her left when cooking, unless she needs to use both. The Ovambo in Namibia will never pass you anything with their left hands and a left-handed greeting is a downright insult. The Wachagga tribe apparently goes so far as to exclude left-handed men from hunting and warfare, since they are believed to bring bad luck. No one knows the origin of the most terrible story of all, which began to circulate in the early twentieth century. There were said to be tribes in Africa that cured a child of left-handedness by burying its left hand in a hole over which they then poured boiling water. The mutilated child had to use its right hand exclusively for the rest of its life.

This is a typical case of a story that demands a large dose of scepticism. Quite apart from the fact that its provenance is hazy, a primitive society that treats the few hands available to it as wastefully as this won't exactly increase its chances of survival. In fact we should be distrustful of much of the information we have about early contact with tribes in inaccessible parts of the world. Its origins often lie with a single missionary, adventurer or anthropologist, so independent confirmation is lacking. Such individuals, especially in colonial times, tended to look through a powerful European-Christian lens and therefore probably misinterpreted much of what they saw. They didn't necessarily speak the language of the people they were observing. All those lonely seekers after fame and fortune, all those enthusiastic servants of the Lord must have guessed, romanticized, exaggerated and even invented a great deal. They undoubtedly sometimes mistook incidental events for customs.

Now and then researchers have had their legs deliberately pulled by the peoples they were studying, and often with good reason. Strangers brought excitement, amusement and intriguing goods and customs. You want to keep on the right side of people like that, in the hope that they'll stay and give you more things. So the newcomers sometimes got more than they bargained for when they displayed an interest in something. It was an early form of the tourist industry. All over the world, 'natives' eagerly threw themselves into traditional handicrafts, dances and musical performances – many of which had never existed before, in that particular form at least.

Whatever the precise truth of the matter, the customs on which the explorers reported often seem to have disappeared without trace by the time the deep interior of Africa and other continents was opened up.

One example of jumping to conclusions is the story of how the Ovimbundu of southern Angola gravely insulted each other with their left arms. It goes like this: you stick your left arm in the air, fist clenched, and with your right hand clasped around your left wrist you shake your left arm back and forth. Observers immediately concluded that the Ovimbundu associated the left with 'very bad'. But consider what the gesture might represent: if the fist is the head of a person you want to insult, then the wrist is his neck. The other hand squeezes the wrist violently, while shaking it to indicate that the person is struggling. Seen in this way it's a vivid gesture meaning: I could strangle you! The fact that the left arm takes the role of the victim is hardly surprising. Like human beings everywhere, most Ovimbundu are right-handed, so the right hand does the strangling. Only indirectly does this have anything to do with the symbolism of left and right.

5
Lovers of the Left

The association of right with good and left with evil can be found all around the globe, but there are occasional exceptions, the main one being China, the oldest of all the world's political entities.

It seems at first sight as if things are no different in the venerable Middle Kingdom from how they are elsewhere. On the shores of the Yellow River, around 90 per cent of the population is right-handed and 10 per cent left-handed. As in other parts of the world, Chinese left-handers have been forced to write and eat with their right hands, but in contrast to Europe and America, this has nothing to do with negative attitudes to the left side in general; in fact, in many ways the left side is traditionally seen in a favourable light. In China left-handers are no less valued than right-handers, but there are traditional reasons why certain actions must be performed with the right hand.

The ancient Chinese world view deals dextrously with the incongruities of the country's history. Western symbolism is based on the contrast between good and evil. Judgement is brought to bear: good things are good and bad things bad by definition. They are opposite poles. The Chinese have always seen matters differently. Rather than being based on a static distinction, Chinese symbolism is all about a balance between the concepts of Yin and Yang. These are opposite poles too in a sense, but they do not directly involve a value judgement. Yin is not good by definition and Yang evil, nor vice versa. They complement each other, creating a harmonious whole.

The symbolic associations attached to Yin and Yang seem suspiciously familiar. Yang is masculine. It stands for leadership, sky, light and the sun. Yin embodies, among other things, the female, submission, blood and earth. But there the parallels end. Astonishingly, Yang is connected with the left, Yin with the right, precisely the opposite of what those of us brought up in Western cultures would expect.

The reason is that the Chinese traditionally orientated themselves towards the south. Emperors, kings and noblemen received their vassals on a podium, facing south as they did so to emphasize their connection with the sun, leadership and Yang. This meant the direction of the rising sun, the east, lay to their left, so the east and the left belong to Yang as well.

The left side is therefore the place of honour, as demonstrated by military tradition. Army leaders in ancient China were archers, commanding their armies from war chariots. Leaders were expected to stand facing south, so an army always had to advance southwards. This was made possible by having a red flag carried at the front of

Three-man chariot team from the terracotta army of Emperor Qin Shi Huang, standing behind their three horses. The wooden wagon and its shaft have rotted away completely, as have the leather reins the man in the centre once held in his hands. To his right is the pikeman, his hand still curled around his long-lost spear, and to his left is the commander, an archer. Since the commander seems to be supporting his weapon the way a modern soldier holds the barrel of his rifle, he was probably equipped with a crossbow.

This bronze two-man war chariot was found in 1980 in Xi'an. Here too the driver of the chariot stands to the right, while the commander's heavy crossbow can be seen at the front left of the chariot and a quiver for cross-bow bolts has been affixed to the left side.

the column, symbolizing the sun. But where on a war chariot does the commander stand?

In 1974 peasants ploughing the fields in the district of Xi'an in the province of Shaanxi chanced upon the 7,000-strong terracotta army of Emperor Qin Shi Huang, which had lain underground for more than twenty-one centuries, ready to fight off any attack on the emperor's grave. As well as infantry and a complete command post, the army had more than a hundred war chariots. The vehicles, made of wood, had rotted long before, but many of their life-sized teams of terracotta soldiers had survived largely intact. It became clear how Chinese three-man chariot teams operated.

In the middle stood the charioteer. To his right was a pikeman, in an ideal position to cause carnage on the right side of the chariot. The commander stood to the driver's left, not the most convenient place for him, since in most cases he was a right-handed archer. His right elbow all too easily got in the way of the driver and it was hard for him to lean over the side and shoot forwards past the horses. But the left side was the Yang side, and because the symbolic south was in front of the char-iot, the left was to the symbolic east, which was also Yang. So even on a two-man chariot with no pikeman, the commander stood to the left of the charioteer.

If left was so favoured, why were left-handed people not the sub-ject of outright devotion? Why are Chinese children always taught to eat and write with their right hands? For an explanation we have to go back to a mythical era, to the story of the Chinese equivalent of Lucifer.

In ancient times, when the world was still young and empty, China was appropriately known as the Middle Kingdom. It was at the centre of the world, and little existed outside it. This was an orderly universe, a balanced and strongly built house. The earth was its floor and directly above was its roof, the sky, with the sun at the centre and around it the stars. The sky was firmly supported on four enormous pillars, one in the north-west, one in the south-east, and one at each of the other two corners. The emperor's throne was in the middle of the universe, which meant that the sun stood directly above the palace and he cast no shadow. From that glittering, sun-drenched throne the emperor ruled his kingdom supported by loyal and competent ministers.

One terrible day one of those ministers, the evil Gong Gong, rose in rebellion against the emperor. He was narrowly defeated, but only after he had severely disturbed the orderly world by breaking into pieces the north-western pillar supporting the sky: Mount Buzhou. The results were dramatic. The loss of that pillar made the sky sink down in the west, and the earth, now that it no longer bore the weight of the sky in the north-west, tilted slightly eastwards. So it came about that the sun and the stars daily slide past from east to west and all the rivers in China flow down from the west to the east, where the seas were created by all that water pouring into them.

Everything had been displaced by the unimaginable violence of the rebellion, the sky towards the west, the earth towards the north-east. So the Imperial Palace never again stood in its original place on the equator, at the centre of the universe.

From then on the world was like a half-collapsed house, with its roof tilted in relation to the floor. In the west the sky stuck out some way beyond the earth; in the east the sky was too short to cover the ground. In other words, on the upper side of the universe there's a disharmonious surplus to the west, while on the ground the disharmony is to the east. Such a state of affairs has far-reaching consequences for a vision of the world in which everything, large or small, ultimately forms a single entity, including the human body, which according to traditional Chinese beliefs is a replica of the universe on a small scale. The round head represents the round sun and the rectangular feet the flat earth. We can tell how seriously this idea was taken from the fact that for a long time dancers at the Chinese court were forbidden to perform handstands, since by doing so they would literally turn the world on its head, the last thing any emperor would ever want to encourage.

That which holds true for the entire universe applies to human beings as well. So people too are imbalanced at the top on the western side. Because of the southward orientation, this means the right side. Lower down they suffer from disharmony on the eastern or left side. Since the boundary between top and bottom lies at the navel, eyes and ears belong to the top half of the body, while feet, legs and hands are part of the lower half. After all, when at rest the hands are at roughly crotch level. Because the imperfection at the top resides on the right, the Chinese regarded the left eye and left ear as superior to their counterparts, but the opposite applied to hands and feet: the right foot and the right hand are superior. The right hand therefore served for precision tasks such as eating and writing. This complicated reasoning clearly shows that although the Chinese may have a philosophical bent, deep down they remain a very practical people. The left side can be favoured yet the right hand preferred – in most cases this suits people nicely.

That the preference for the right hand has a practical rather than a moral basis is demonstrated by the ancient Chinese version of the victory procession. A victorious general would hold a sword in his right hand and a flute in his left. War is manual labour, and bloody and unhappy labour at that, of the kind that brings misfortune. War is Yin. The right hand – also Yin – is allocated this particular kind of work and valued accordingly, but a more important role is reserved for the left hand, which is Yang. Victory processions mark war's end, indeed its successful conclusion. They signify that peace has arrived, so it's the left hand that carries the symbol of peace, the musical, non-violent flute that brings pleasure and repose.

There are other places besides China where the left hand is not necessarily regarded in a negative light. In East Africa, for example, a tribe called the Wageia-Kavirondo regards the left as the side of good fortune and the right as the side of misfortune. Anyone who stubs his right foot twice while planning to leave on a journey is advised to stay home. If he stubs the other foot, the journey is sure to be a success. The Wachagga, whom we met earlier as the tribe that excludes left-handers from hunting and warfare, are utterly realistic when it comes to feet. If a Wachagga stubs his left foot during a journey, then it does not augur well, since the right foot is seen as the lucky foot, but if he carries on anyway and returns after a successful trip, this is seen as proof that his lucky foot is his left.

Among the Masai, the famous herdsmen of the East African plains, there's a superstition linking the left hand with health – something that is quite common elsewhere too. On the night of the new moon the Masai cast away sticks or stones with their left hands, calling out 'give me a long life' or 'give me strength' as they do so. Unusual colour symbolism is involved here. Most peoples in Africa associate right with light and white, left with dark and red, just as Pythagoras did in his Table of Opposites. For the Masai precisely the opposite applies: a man who has killed an opponent in battle is permitted to paint his body, the left side white and the right side red.

Even in Judeo-Christian thinking, left-handedness is by no means always saddled with negative connotations. The Bible certainly creates an impression of bias towards the right, but that's mainly because the right hand and the right side are mentioned so often, and because on the Day of Judgment the damned move away to the left towards hell. Left-handedness is mentioned only twice explicitly, on both occasions in the Book of Judges, and both times in a strikingly positive sense. Ironically, both occurrences concern the tribe of Benjamin, a name that literally means 'sons of the south' or 'sons of the right hand'. The tribe got its name from the fact that its tribal area, the region around Jerusalem and Jericho, was directly to the south of the territory of the Ephraim, the dominant tribe, whereas the ancient Israelites, as was usual in the Middle East, oriented themselves towards the east.

One of the two stories concerns a gang rape by a group of Benjamites, which results in a full-scale civil war between the Benjamin and other tribes. The bloody battle goes badly for the tribe of Benjamin, but not before the vengeful attackers have suffered heavy losses. The relatively small but courageous army of the Benjamites deploys a unit so unusual that it receives special mention: an elite battalion of 700 left-handed slingmen. This seems a huge number for a small people like the tribe of Benjamin; we should probably take 700 simply to mean 'an awful lot'.

The other story concerns the left-handed hero Ehud, who ends the occupation of Israel by the Moabites. Ehud is specifically chosen by Yahweh to assassinate Eglon, the Moabite king. It is no easy task, since Eglon is as tightly surrounded by security men as a modern American president. Ehud hides a knife under his clothes, against his thigh, and accompanies a delegation to an audience with Eglon. He manages to get past the bodyguards. Like everyone else, he is body-searched, but ineffectually, since the guards don't expect to find anything on the inside of his right leg. Normal people, which is to say right-handers, would carry a weapon on the inside of the left leg. That was precisely why Yahweh had chosen a left-handed man for the job. So Eglon does not survive the visit, whereas Ehud, in true James Bond style, escapes over the rooftops and becomes a national hero.

Both stories suggest at the very least a pragmatism that many left-handers will recognize from their own experience. The overwhelming majority of their fellow human beings regard the left hand as the bad hand, but sometimes left-handed people can be useful, and why wouldn't you want to make use of them? It seems the negative associations surrounding the left are not insurmountable, even in biblical mythology. This raises the question of how strong the connection between the left and evil in our symbolic thought patterns really is, and whether perhaps there are other connections that make for rather more subtle attitudes towards the left.

6

Magic and Superstition

What are the norms, values and value judgements by which people are guided in daily life, in 'thinking without thinking'? They emerge vividly from superstition, which is closely bound up with our deepest being and survives in the face of any amount of oppression. Early church patriarch St Augustine fulminated against it, as has virtually every authority since, but that hasn't helped. Even people with apparently modern, businesslike, rational attitudes can be found driving around with horseshoes bolted to their radiator grilles or trying to avoid walking under ladders. More importantly still, superstition is informal. It arises spontaneously, rather than being devised and prescribed, and nothing is carved in stone; there exist no authorized versions. It therefore gives us a fairly accurate impression of the things we truly associate, deep down, with left and right. At least, it does so a good deal more directly than the simplified and to some extent artificial systems of etiquette, philosophy and established religion that are based on it, such as Pythagoras' Table of Opposites.

Even just within Europe, superstitions that refer to the left side, the right side and the left hand are plentiful. Many have to do with witchcraft and duplicity. To start with the latter, there are many traditional ways of committing perjury with impunity. In some places it's enough to keep your left hand in your pocket while taking the oath, or to touch your jacket button with it as you give your word of honour. We can guard against this particular form of duplicity by making it compulsory to raise the left hand while swearing to tell the truth and nothing but the truth, but in a number of cultures this in itself is reason enough to regard an oath as non-binding – in America, for example, as viewers of its countless courtroom dramas will know.

From present-day Belarus and surrounding regions comes a belief in the *hecktaler*, a coin that always returns to its owner as long as he pays it out with his left hand while gently treading on the seller's left foot. This sounds ideal for anyone hoping to make a quick buck, but how do you come by such a coin? *Hecktalers* were originally owned exclusively by people who had sold their souls to the Devil, or by Jews – it seems people in Belarus made no firm distinction between the two. Others could make a *hecktaler* their own by treating a buyer who offered to pay with one in a similar manner, in other words by accepting the coin with the left hand and surreptitiously treading on the buyer's left foot.

Anyone unable to lay his hands on a *hecktaler* could always get rich by gambling, since good luck could be guaranteed by the following elaborate, if unsavoury, procedure. The day before a gambler has a chance of winning the jackpot, he catches a toad. That evening he pushes a needle and thread in through one eye of the toad and out through the other, right through its head. The story doesn't mention whether the poor creature has to be alive. Then the toad is tied to the fingers of the left hand by the thread, where it stays until the following morning. The gambler is certain to have luck on his side.

The left–right distinction is often a key element in procedures for overcoming a fear of supernatural figures such as witches, dwarves and vampires. For example, both right and left are of crucial importance in a centuries-old remedy for young parents fearful of evil dwarves intent on replacing their child with a changeling. Lay a right shirt sleeve and a left sock in the cot next to your child and it won't be stolen. You'll just have to hope it doesn't suffocate amid all those bits of fabric. From time immemorial, bakers in the German region of Swabia have pressed the fingertips of their left hands into the dough of the last loaf to go into the oven to prevent witches from gaining power over the bread. And did you know that a vampire in its coffin can be recognized not only by the fact that the body does not decompose, retaining a healthy complexion, but because the left eye always remains open? Now that we're in Romania, the land of Dracula: if a person in that part of the world bleeds from the left nostril, then someone in his or her family has just died. Another bad omen is the barking and wailing of dogs, a harbinger of fire, death and war. Calamities that announce themselves in this way can be averted, it is said, by holding out to the canine in question, with the left hand, the heart of a black dog with a dog's tooth stuck through it.

All this brings us to another connection, this time between the left hand and the act of killing. If, for example, a witch has turned herself into a toad, then she can be killed only with an axe held in the left hand.

Ridding yourself of disagreeable acquaintances is rather more difficult, but here is a somewhat impractical tip from Central Europe: furtively acquire some blood from your prospective victim and wipe it over the sole of the left foot of a fresh corpse just before burial. Your acquaintance will gradually waste away and die.

Then there is love, of course, which prompts the strangest antics of all. In the region of Landshut in Germany, people once believed that a young man out courting could win the love of a girl by secretly stealing a blouse she had worn while cleaning the house, then urinating through the right sleeve. If she later proved a disappointment, the flames of passion could be extinguished by using the left sleeve as a urinal. Fairly revolting, but as nothing compared to the courting advice offered by people in the region around Breslau and Wroclaw. The admirer of a girl he wishes to marry must start by swallowing a whole nutmeg, which is indigestible and will emerge after a day or two in practically the same condition as it went down. It must be fished out of the ordure and rinsed clean. The next Friday the aspiring young man needs to clamp it in his left armpit for an hour 'in Hora Veneris', or 'at the hour of Venus', in other words at sunrise. The nutmeg is then grated and the lady in question invited for a meal. He must take the opportunity to feed the entire nutmeg to her, after which she will be putty in his hands. Some say it works with cattle too, although one wonders what anyone would want with a hopelessly lovesick cow.

Central Europe is not the only region where the left side of the body is of relevance to supernatural means of facilitating courtship. A third-century Egyptian papyrus describing magical rites includes a recipe for a love potion in which the essential ingredient is blood from the left ring finger.

Many of these stories have unpleasant aspects, usually a cunning deception or the temptation of innocents, but some of the superstitions surrounding the left side and the left hand seem unrelated to either good or evil. The widespread idea that small children who have never looked in a mirror can see themselves reflected in their left hands is simply enchanting. What should we make of the trick that Swiss milkmaids believed would make a recalcitrant cow stand up to be milked: you only needed to wind your left garter around the animal's right horn. The left hand, perhaps surprisingly, is capable of driving out evil, and not only in the form of a Swabian witch or an enchanted toad. Wiping the left hand over the bedclothes will prevent nightmares, for example. We find something similar in the Punjab in northern India, where a young man will wind an old rag around his left arm to ward off the evil eye.

Finally, here is a truly bizarre belief in which the left can be both good and bad. Not so very long ago in Western Europe, the drawing of lots determined who would be drafted into the army. If you were rich you could simply hire a replacement to endure the barbaric living conditions in the barracks on your behalf, but for men of few means the lottery was a true sword of Damocles. The drop in a family's income entailed by having to manage without a son who had just reached adulthood might be a devastating blow. For such people the only imaginable solution lay in magic and superstition. It was believed that a young man could rely on meeting with a favourable result as long as he did everything with his left hand for three days before lots were drawn, including crossing himself, assuming he was a Catholic. He also had to make sure he drew his straw with the left hand. Whether this particular left-handed scheme is good or evil depends on your point of view. A militant patriot with plenty of money will regard it as a cowardly means of avoiding an honourable duty, whereas an impoverished wretch is more likely to see it as a god-given opportunity to protect his family from hunger and want.

One thing will by now have begun to emerge from this array of stories: the left side of the body has magical, enchanting properties that can be used to do good as well as harm. The following nuggets of wisdom show that the magic involved is usually of a highly specific kind.

How can a persistent fever be treated? Here's a method cheaper than a visit to a doctor: the sufferer winds a piece of blue wool around a toe on his left foot and leaves it there for nine days. On the tenth day, before sunrise and without saying anything, he goes to an elderberry bush, takes the wool from his foot and winds it around a branch. The fever lifts. Similar methods were used to treat innumerable other ailments. The length of time the wool has to be left in place, the precise body part, and the kind of bush or tree that would absorb the sickness all vary, but a left finger, arm, leg or toe was always involved. A bad coughing fit in the Tirol? Let your left arm hang loose. Persistent nosebleed? A thread wound tightly around the little finger of the left hand does wonders. Infants could be protected from gout by three drops of blood from the left ear of a black sheep. Sick pigs are curable too, the ancient Romans believed: before sunrise, dig up the roots of a hellebore with your left hand, drill holes in the sick animals' ears and push the roots through. The creatures will revive even as you watch.

Conkers relieve all kinds of complaints, from toothache to rheumatism, as long as you keep them in your left trouser pocket – although there are some regions where people believe they should be put in the right pocket. Medicinal or magical plants can be picked at many different times – just before sunrise or perhaps at midnight, during the new or the full moon – but dozens of cures and spells agree that it must be done with the left hand, preferably with the thumb and ring finger.

All this suggests that the magic of the left hand is the magic of life and death, of sickness and health, including the healthy sickness we call love. This is the most important theme, one that recurs over and over again – far more so than trickery and deceit. Even official Catholic symbolism, which generally doesn't seem to regard anything about the left side and the left hand in a positive light, has at least one incontrovertible rule that confirms the left's connection with health, love and vitality. It concerns the marriage ceremony, in which the bride and bridegroom place rings on the ring fingers of each other's left hands. This custom is derived directly from an ancient Roman engagement ritual,* a contract sealed with a ring on the left ring finger, the *digitis medicinalis*, which was seen as having a powerful influence on health. The ring afforded protection, and preserved love. Isidore of Seville wrote that people of his time believed an artery ran from the left ring finger straight to the heart. Perhaps that's why the blood of this particular finger was required for the libido-boosting drink described on that Egyptian conjuring papyrus.

Another indication that European Christian civilization is thoroughly permeated by a belief in the connection between the left hand and health and vitality can be found on the ceiling of the Sistine Chapel in Rome, in the famous scene where God gives life to Adam. Naturally he does so with his right hand, but Adam, created a moment earlier in the image of God, accepts life with his left. This is remarkable, since left-handedness was generally seen as a characteristic of the Devil. An often heard explanation is that Michelangelo positioned Adam as he did because any other arrangement would have messed up the composition. That seems highly improbable. A pope who hires the best painter in the world to create a work of art at such a holy place wants his money's worth. He surely can't be palmed off with talk of the job being too difficult – and Julius II was hardly the easiest of popes.

There's a quite different, well-documented, and for the church identifiable and acceptable reason for Adam's left-handedness: the left hand

* The church engagement ceremony we know today, an extension of the marriage ceremony with the rings on the right hand, is a later innovation.

Adam receives life through his left index finger. Michelangelo, Sistine Chapel, Rome.

represents life itself. We see this reflected in another panel of the same section of the ceiling, where Adam and Eve are thrown out of paradise. The fateful deed is depicted on the left: the eating of the forbidden fruit. On the right the consequences are shown. Two pitiful, remorseful and utterly despairing figures flee the implacable Angel of Death who floats above them. This angel is left-handed – and not, of course, because he's meant to be a henchman of the devil, since he's acting on God's orders. No, his left-handedness indicates that this is a matter of life and death.

Superstitions about deceitfulness, such as those ways of swearing a false oath and the use of the *hecktaler*, have nothing do to with the complex of sickness and heath. They relate to a quite different theme: inversion. It's a subject that must be almost as old as mankind and it flows directly from our tendency to polarize and to split things in two. In the Judeo-Christian world it is also directly connected with the distinction between good and evil, since here we encounter the inverse of God: the Devil. God looks like man, so he's naturally right-handed as most of us are, but the Devil is left-handed, a reversal reminiscent of the way he demands to be kissed not on the cheek or the mouth but on the buttocks or anus.

Adam and Eve eat fruit from the tree of knowledge and are expelled from paradise. Michelangelo, Sistine Chapel, Rome.

Inversion as a characteristic of the Devil is alive and well to this day, and not just in the inverted symbols and rituals of the black mass. As recently as 2008 the Dutch Christian fundamentalist Minister for Youth and the Family, André Rouvoet, publicly declared himself proud of the fact that as a teenager he'd investigated whether pop and rock records contained messages from the Devil by playing them backwards. If he found one that did, he would break the guilty record in two – quite an effort in the days of vinyl. He insists he would do the same today.

The theme of inversion can be found the world over, but it's by no means always a matter of good and evil. In the ancient world, sacrifices of fish or small animals to the ancient earth gods were often made using the left hand, with the creature's head downwards. Here inversion is merely a logical consequence of the notion that the gods of heaven live above us and the earth gods beneath.

Among the Toraja on the Indonesian island of Sulawesi, inversion is closely connected with the distinction between life and death. The right hand is for normal life and the living. The Toraja therefore care for the graves of their ancestors exclusively with their left hands, and even a handful of rice sacrificed to the dead must be scattered with the left. They imagine the dead as white, whereas their own skin is dark. Left-handedness is unacceptable in their culture, not because the left hand is bad but simply because, shockingly, a left-hander treats other people as if they belonged to the realm of death. Offering someone a drink with the left hand strikes the Toraja as like offering someone a wreath as a birthday gift.

The left and right halves of our bodies, then, are clearly and fairly consistently connected with two themes in folklore. One is the magic of sickness and health, which resides on the left. The other is the motif of inversion. But when superstitions concern things outside ourselves and how they move around, then at first sight they make no sense. Arbitrariness reigns, we might think. In Central Europe sheep crossing from right to left are a sign of misfortune, whereas both the Romans and the native Americans thought that crows heralded death and destruction if you watched them pass from left to right. The French agree that crows are a bad omen, but they remain convinced that if the birds approach from the left then their evil effects can be avoided, so only crows coming from the right are worth worrying about. A cuckoo calling off to your left means bad luck, but to your right it's a sign of good luck. In some places people believe that sheep or pigs bring good fortune as long as they pass on the left side. If they pass by to the right you need to watch out. In other places people are just as convinced that the opposite applies.

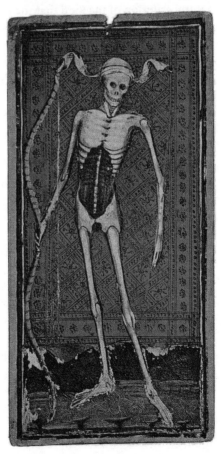

The Tarot deck's Death carries his bow in his right hand since – and this is surely no coincidence – he is left-handed.

Official church rituals are no less muddled. Catholics make the sign of the cross by touching first the left shoulder, then the right, but Anglicans and the Eastern Orthodox do it the other way around – although with the right hand; no difference there. This probably has something to do with the fact that both groups of dissenters from the mother church wanted to differentiate themselves as far as possible from their former brothers in the faith. This is a prime example of inversion. The strangest case of inconsistency in direction concerns the ancient Romans, fervent believers in all kinds of omens that had to do with the flight of birds and natural events of that sort. The left was originally regarded as lucky, but under the influence of Greek culture the Romans seamlessly adopted the belief that the right was the favourable side. Fate, it seems, could take its cue from fashion.

The cause of all this confusion might perhaps be the absence of any natural criterion of the kind that the axis of our bodies represents in relation to our body parts. Left is almost always associated with negative things because of the inversion motif, but what does left mean in the world around us? That depends whether you concentrate on where something is coming from or where it's going to. In the former case, animals we hear to our left, that pass us to the left or cross our path from left to right may be bad omens. But if we look at their direction of movement, everything is reversed and suddenly whatever comes from the right is bad.

With circular motion too, the various beliefs look chaotic at first, but appearances can be deceptive. Here a third theme lies waiting to be discovered.

A wide range of ritual movements are performed clockwise. The processions held by Catholics always move around a church clockwise, just as a priest performing Mass moves clockwise around the altar. When new houses are blessed in today's Europe, people process around them clockwise. In party games we take turns in clockwise rotation and cards are always dealt out that way round.

Clockwise motion can be seen as rightward in direction, so it might seem as if the connection between the right and favourableness is responsible here once again, but in the early twentieth century French sociologist Robert Hertz came up with a different explanation. He claimed that circular rituals were intended to reinforce a common bond, a sense of 'them and us'. People turn their right shoulders towards the safe centre of the group and their left shoulders towards the hostile outside world.

As a result, they automatically circle clockwise. Neither explanation can be right. Hertz does not make clear, for example, why we would choose to turn our right shoulders inwards. Surely a right-handed majority would want to use their right hands to defend themselves against a hostile outside world. Furthermore, neither Hertz's hypothesis of cosiness nor the connection between clockwise rotation and goodness and good fortune explains why a number of traditional rituals involve anticlockwise motion.

The sports world offers one highly visible example. Almost all racetracks, whether for dogs, horses, people or cars, are used anticlockwise. The propellers of planes and helicopters turn anticlockwise too, as do windmills.

The best explanation for these phenomena presented so far lies in the path described by the sun. In the northern hemisphere the sun moves clockwise across the sky. Although they may be completely unaware of it, priests and believers who circle churches, altars or houses are following ancient heathen sun rites. They are reflecting the course of the sun as it passes around the place at the centre of the ritual: the church, the altar or the house to be blessed. Precisely the same rituals exist in the southern hemisphere, but this can be explained by the fact that all today's dominant cultures arose in the northern hemisphere. It would be wonderful to know whether the Incas of South America or the long-lost Bantu cultures of the Congo and southern Africa had circular rituals long before the first European explorers set sail, but they all disappeared long ago, leaving no evidence as to their directional sensitivities.

The traditions of board and card games adhere to exactly the same principle. The players represent the path of the sun, the 'turn' is the sun itself and the point around which the sun revolves is the centre of the board or the gaming table. Each player watches the course of the game pass on the other side of the table in the direction of the sun, from sunrise on his left to sunset on his right.

So what about windmills? They may appear to turn anticlockwise, but that's because we tend to look at them from the wrong side. The reference point for a windmill is not the chance passer-by, looking at the mill from the front, but the miller himself. He is generally inside, behind the turning sails; from his point of view they turn in the same direction as the clock and the sun.

The crankshaft and therefore the drive shaft of an internal combustion engine, as well as aeroplane propellers, simply follow the tradition that arose centuries ago when the first windmills were built. The engine in the earliest motor vehicles quickly came to occupy the place formerly taken by the horse, in front of the driver. Seen from his perspective, engines and propellers turn clockwise. The demands of industrial production soon established an unswerving standardization that was more enduring than could ever have been created by superstition.

Racetracks, finally, differ from all other circuits in the position of their reference point, which lies outside them. The crowd in a stadium or at a racetrack does not usually sit in the centre of the arena; in fact, the centre tends to be virtually empty. Instead people look on from outside. In contrast to players of board and card games, they are not participants but literally outsiders. The action they watch most closely takes place on the nearer side of the track. If the movement observed by the crowd were to circle clockwise, then the competitors would be seen to move from right to left, passing the spectators in the opposite direction to the sun. By using the track the other way round, the usual effect is preserved; the participants pass, like the sun, from left to right.

So all circular movements conform to the same principle, following the sun. The reference point is all that differs from case to case. With processions and games it lies in the thing the circuit revolves around, with windmills it lies inside the mill, with engines in the position of the driver, and at racetracks in the eye of the beholder, outside the circuit.

7

The True Nature of Left and Right

Inversion, the path of the sun and healing magic: these are the three contexts in which the left side and the left hand occur in modern-day symbolism. Only the last of the three is interesting in its own right. With inversion, the choice of the left hand is simply derived from the fact that right-handedness is the norm, and rituals inspired by the path of the sun across the sky are dictated by the fact that the sun happens to travel from left to right as seen from the dominant northern hemisphere. The left is of essential importance only in magic. The symbolic link between the left side of the body, the left hand in particular, and health and sickness, life and death, is not derived from anything else, at least not directly. It has to do purely with the opposition between left and right. This is a puzzle, since although left and right are opposites, it's less clear what the opposite of health and sickness could be, or of life and death.

Of course with matters of this sort it's impossible to prove anything beyond doubt. We are dealing with soft values – irrational feelings, intuitive judgements, things that lend themselves poorly to analysis – and we cannot ask our ancestors to tell us about their motives. We can't even directly observe the foundations on which we construct our everyday, subconscious view of the world, but we can formulate a reasonable conclusion about what we suspect is going on with all that health magic and its link to the left side of the body. First we need to return to classical antiquity.

More than 2,000 years ago, in the Hellenic period, a goddess with Egyptian origins called Isis developed into a divinity that was popular all the way from England to Mesopotamia. Isis was a manifestly female figure who symbolized the elongated land called Egypt as it waited to be fertilized by the flooding of the Nile, a river represented by her husband

Osiris. Together the couple symbolized the cycle of living and dying, the idea that life returns after death, just as new life sprouts from the parched earth every spring. In that flourishing period the Isis cult grew into a mystical religion that was in some ways similar to the later Christian faith. It had initiation rites that included baptism, and people who underwent them were literally expected to see the light. Those who had been initiated would live on in the Elysian Fields after their deaths, under the protection of Isis, on condition that they had fulfilled the duties laid down, one of which was permanent chastity. The Isis cult was centred on values familiar to us, including self-control, asceticism and a sense of guilt and remorse. The Hellenic world, it seems, was becoming ripe for a religion like Christianity.

Processions were held for Isis, and those taking part carried a wide range of symbols. One of them, Apuleius tells us in his *Metamorphoses*, was the image of a left hand as a symbol of justice. He adds that it was highly appropriate, since the left hand's innate clumsiness made it better able than the right to symbolize that particular virtue. This indicates exactly what the left hand portrayed: not artful, juridical justice, with its manipulations and complex, logical interpretations, but a sense of justice, the emotional experience of justice, the feeling of fairness and just deserts.

Remarkably, we find something very similar in Judaism. According to ancient Jewish tradition, Yahweh holds mercy and the Torah in his right hand, life and justice in his left. Not only is the left hand once again associated with life, but it also symbolizes justice, as in the Isis cult, except that this time the opposite pole is the Torah, the written law. Justice must therefore be understood here too as a purely emotional conviction that right has prevailed, that things are as they should be, rather than the confidence that certain rules or laws have been respected. Anyone who has received a parking ticket or been thrown out of a school class on dubious grounds knows that there can be a considerable difference between the two.

Isn't this contradicted by the fact that the right hand of that same god holds mercy as well? Surely mercy is an emotional business? Not at all. Our intuitive sense of justice has to do with atonement, with a settling of accounts, with revenge. Ultimately it means an eye for an eye and a tooth for a tooth, as demonstrated by sayings such as 'as you sow, so shall you reap', 'on your own head be it' and 'good comes to those who do good'. It's brutish and unrestrained, with no room for empathy or for allowing merciful compassion to serve as justice.

Mercy by contrast puts a civilized, rational brake on our desire for retribution. We might want to tear those who wrong us limb from limb, to hang, draw and quarter them, to tear out their tongues, but instead, entirely rationally, we muster some understanding for their motives and circumstances, consider what the further consequences of vengeance would be and satisfy ourselves with less than our due. Mercy has nothing to do with lovingly stroking the head of a naughty child – that's simply tenderness or pampering. Showing mercy is a way to interrupt a spiral of violence before it gets out of hand, mainly on rational grounds.

Here the true significance of the contrast between left and right emerges at last: emotion as opposed to reason, things felt but not understood as against analysis and knowledge. Left also signifies the things that inevitably happen to us in contrast to the things we control, magic as opposed to skill, and this fits perfectly with its link to health and sickness, life and death. If there's anything mysterious and unfathomable, anything with a magical charge, then it's the business of sickness and health. This remains the case today, to judge by the popularity of herbal lore, faith healers and the laying on of hands. And let's not forget the placebo effect.

Elsewhere too we come upon indications that the contrast between the rational and the irrational represents the ultimate significance of the symbolic contrast between right and left. We need only look at advertising, especially on television. All over the world, left is associated with femininity, right with masculinity. We saw this in Pythagoras' Table of Opposites, but it emerges in countless other ways too. Think for example of the belief that sperm from the left testicle produces girls and sperm from the right boys. It's a superstition that goes back to another early Greek philosopher, Anaxagoras, and it survived well beyond the Middle Ages. Over the centuries, a great many men have tied off their left testicle in the hope of conceiving a son.

Even on the other side of the world, people think the same way about male and female. The Maori of New Zealand, to take one example, call the right *tama tane*: the male side. *Tama tane* is also the term for the male sexual urge, for power, for creativity and for the east. The female side, *tama wahine*, has precisely the opposite connotations. Many Bantu tribes in Africa regard the right hand as the strong, masculine hand and the left hand as feminine and weak.

In contemporary advertisements for fashionable luxuries, such as certain drinks, or night clubs and fashion items, the man is generally the smooth, self-assured type who is very much in control. He goes about either in casual clothes of calculated nonchalance or in sharp suits. He drives a classic car of the kind available only to true go-getters, expertly sails a seagoing yacht single-handed, stripped to the waist, or carries a briefcase which clearly contains extremely important documents, indicating that he bears immense responsibility. He's a man of the world, exuding expertise and self-confidence, a good sport but clearly someone who sets his own boundaries and upholds them.

Women in these commercials are of an altogether different nature. No longer the dumb blondes or gentle, self-sacrificing mother figures of decades past, they take the initiative and dress in tantalizing, shiny lingerie or enchanting but utterly impractical outfits. Sometimes, playfully, they may wear men's clothing, but in that case they're redolent of naughtiness with a hint of irresponsibility. Clearly such a woman couldn't care less about any conventions or boundaries that prove inconvenient. Sometimes, rippling like a tiger, she stalks a man, tempting him into a wild tango. She may even slink towards him across a crowded bar, oozing an aggressive, mysterious sensuality. He goes along with her as far as it suits him, even a fraction further perhaps. She may get to the point of tearing the clothes from his body, although only to make herself seem all the more tempting, or to lead him into a yet more irresponsible game. She goes to extremes, behaves with undisguised hostility towards other equally ravishing women, and regards life as a feast for the senses. If she has ever worked at all, then she probably had something to do with painting or sculpture.

Among the prime examples of this division of roles are the advertising campaigns a Dutch gents' clothing company called Van Gils has been running since around 1985. The earliest of its adverts each featured a man who in one way or another fell under the spell of a woman. Once, only half covered by a sheet, he was taken in hand by a heart-stoppingly pretty masseuse, his suit hanging over a chair just out of reach. Another time a woman seized possession of his suit while he was in the bathroom. Each time the tagline went: 'Back in control very soon.' Here we have all the symbolic elements in an undiluted form. Neither the man nor the woman lack courage, but in the woman's case it's the courage to let herself go, to allow herself to be led by playfulness and sensuality, whereas in the man's case it's the courage to take a calculated risk; he delivers himself to the woman, but not completely. The tagline represents what he's thinking. A quarter-century later, in 2009, the Van Gils campaign featured a love-making session in an expensive hotel room. The woman is as playful and sensual as ever, elegantly revealed in her black lingerie. The man is wearing his suit this time, and he has the head and hands of an unpainted mannequin. He plays the game, but without baring an inch.

It's clear what's going on here. Whether or not we agree with the premise on which they're based, these adverts are caricatures of deeply rooted associations: the man stands for control and calculation, in short for reason, while the woman represents irrationality, spontaneity and uninhibited emotion. As Camille Paglia would put it, the woman is the natural, irrepressible element, contrasted with the orderly, controlled, cultured world of the man. If we take account of the established links between femininity and the left, masculinity and the right, then everything falls neatly into place. Left and right are symbols of nature versus culture, magic versus expertise, overwhelming emotion versus controlled deliberation.

8

Strange Creatures in the Uncanny Valley

Nature and culture are as inseparable as Yin and Yang, those two equally indispensable cornerstones of our existence. So how did the left acquire such negative connotations? The answer is fairly obvious. A certain degree of threat inevitably emanates from any kind of magic, because it's unpredictable and unfathomable, and because its effects are not always pleasant. Perhaps more importantly, people readily attribute unexpected, unwelcome events to acts by higher powers, if only as a way of escaping the question of whether they could have, perhaps ought to have done something to prevent them from happening. It's tempting to blame a divinity, a spirit or a conspiracy. Modern Westerners are not as likely as they once were to see a witch, goblin or demon behind all their troubles or setbacks, but even today, when faced with disaster, many people sigh that the Lord works in mysterious ways.

Fear of calamity is never far off in matters of sickness and health, or life and death. To most of us, thinking about health means little more than hoping we won't be struck down by disease. Even our modern, apparently positive attempts to eat well and live healthily have a negative undercurrent, namely the effort to keep sickness, incapacity and death at bay. So there's nothing strange in the fact that a slightly negative odour accompanies the concept of the left, bound up as it is with health and mortality. Left-handers are naturally affected by that whiff of negativity, which rubs off on them to some small degree, and in any case they're relatively exceptional. They constitute a minority, and minorities are always a bit suspect. Not dramatically so, perhaps, but still.

Fear and suspicion of minorities of every imaginable kind is so ubiquitous and timeless that it seems innate. Interestingly, there is some substance to this idea. Evidence shows that our suspicion of everything that diverges even slightly from the norm has a solid biological basis.

In the early 1980s computer technology was still very much in its infancy. Nevertheless the hugely ambitious research community in Japan was captivated by the dream of creating an artificial person, indistinguishable from the real thing: the ultimate electronic butler, human in every respect but without human failings. That dream turned out to be a good deal harder to realize than anyone thought at the time, but we can certainly understand how it arose. Generally speaking we feel most comfortable when our non-human companions seem as much like us as possible. Cats, with their large eyes facing forwards and their little snub noses, look much like small human children. Many types of dog have been bred to develop more or less human features, with round eyes in the front of their heads if at all possible and short muzzles. It's harder for us to warm to cows and pigs, goats and sheep, but they too arouse at least a modicum of tenderness in us. The less an animal resembles a human, the quicker we are to think it troublesome, dirty or creepy. Rats, reptiles, spiders and insects tend not to have particularly good reputations.

The same holds true for non-living companions. Teddy bears and other stuffed animals often look more like human infants than their wild flesh-and-blood equivalents. The prettiest dolls are always the ones with large childlike eyes, preferably the kind that open and close, especially if they can say 'mummy'. Towards the end of the twentieth century the success of the Tamagotchi 'computer pet' proved that people could spontaneously fall in love even with a dull plastic egg that did nothing but clumsily mimic the demanding behaviour of an infant. Apparently we find anything that acts like a human child irresistible, even if it's fretful and unpleasant.

In the earliest phase of these developments, around 1980, one of the Japanese robot pioneers, a man called Masahiro Mori, struggled with exactly that question: in what ways would a robot have to resemble a human being in order to put us at ease and make us feel positive towards it, and which aspects of the resemblance were less important? He made a remarkable discovery. Contrary to what everyone had assumed, people did not feel increasingly at home with artificial human figures the more they looked like real people. As long as the differences were fairly large that assumption did hold true. We feel more affinity for a robot that has a vaguely human shape – think of classic figures like the tin man in *The Wizard of Oz* or the tub-shaped R2D2 in the *Star Wars* films – than for the cold, purely functional sort of industrial robot found in car factories. At the other end of the spectrum, we could easily trust a robot that appeared identical to a human, but in between lies an area that has come to be known as the uncanny valley: if things look very much like us, we feel extremely uncomfortable in their presence. We respond with feelings of intense distrust, fear and revulsion, and the effect is considerably reinforced, Mori discovered, if the thing in question starts to move under its own steam.

In 2008 the Japanese robot industry exhibited some examples of what it could do. The world's media relayed images of lifelike, mostly subservient female figures with supple skin that looked natural when they moved. They were capable of a wide range of realistic facial mimicry, their arms and hands moved the way you'd expect of a friendly employee, and their voices were honey-sweet and convincingly human. Yet something was missing. It was all just a bit too slick, just a little too, well, artificial. Many visitors to the show found themselves involuntarily trembling.

Mori had discovered something of whose existence we've been aware unconsciously for many years. It may explain why foreign actors have been so much in demand in Hollywood as baddies. Men like Max von Sydow and Rutger Hauer made a fortune as a result, and of the 22 super-villains that appeared in James Bond films between 1962 and 2009 only four were born and bred in the United States. Apart from Joseph Wiseman, a Canadian who played the part of the secretive oriental Dr No, all were naturally endowed with thick foreign accents that sounded deliciously scary to Americans: nine Brits, one and a half Frenchmen, two Germans, an Austrian, a Dutchman, an Italian and a Dane. Often

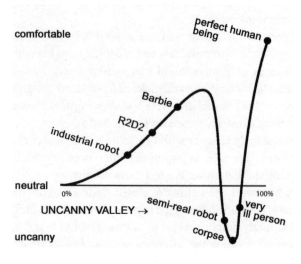

The uncanny valley effect. The horizontal axis indicates the degree of likeness to a real person, while the vertical gives the emotional response evoked.

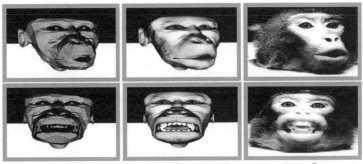

unreal   semi real   real

In the Princeton experiment, Java macaques had no great difficulty with the non-realistic faces, shown to them in grey with red pupils. Nor were they alarmed by the true-to-life colour photographs on the right. But they responded to the caricatures in the centre, which were also rendered in realistic colours, by quickly looking away as if frightened.

they were supplied with an unpleasant wart or other facial disfigurement to make them even more repulsive, or a horrific bloody scar around the left eye like Mads Mikkelsen in *Casino Royale* (2006), emphasizing their status as almost but not quite completely normal people.

Speaking oddly or having a blotchy, spotty or otherwise unpleasant appearance makes you seem slightly ill. It's probably this that's responsible for the uncanny valley phenomenon. We all seem to display an instinctive aversion to fellow creatures who patently have something wrong with them, however slight, whether they are sick individuals you'd do well to avoid or corpses, to which it's always best to give a wide berth. However obvious this may seem, the precise nature of the phenomenon was for a long time unclear. Was it the product of a biological trait, moulded by evolution? Or was it simply a matter of poorly understood cultural conventions?

In September 2009 researchers at Princeton University at long last produced powerful evidence that the uncanny valley phenomenon is indeed a biologically determined characteristic. They showed that it exists in the Java macaque, an animal often used in laboratory experiments. When the apes were presented with pictures that bore a poor resemblance to other monkeys of the same species they responded with mild interest, and they also coped perfectly happily with good-quality photographs, but when they were shown distorted images of their fellow macaques they looked away nervously. They didn't like the look of those freaks at all. The experiment ruled out the possibility of a cultural origin for the uncanny valley; in fact, the scientists concluded it must have existed for aeons, since we know the ancestors of human beings and those of the macaque family diverged at least thirty million years ago.

All things considered, it looks very much as if the uncanny valley phenomenon is crucial to the way we determine whether or not a person is 'one of us'. Creatures that are clearly different from humans have little to fear. We accept them as belonging to a different species. Other animals respond in exactly the same way; nature is a great drama in which everyone tries to ignore everyone else as far as possible. Animals pay attention to other species only if their own interests are at stake. Predators watch out for suitable prey – if they're hungry, that is, and not otherwise. Ducks quarrel, as do coots, but a coot will rarely bother a duck or vice versa, even though the two species live in close company with each other.

Creatures that look like us and behave the same way, that sound and smell like we do, are members of our own species and therefore 'one of us'. But that's not to say we're all alike. As in George Orwell's *Animal Farm*, 'all animals are equal, but some animals are more equal than others'. First of all, we distinguish between our own group and the rest. Fellow group members have a right to our support and loyalty, to friendship and assistance where appropriate. All others are theoretically competitors, who must stay outside what we regard as our territory. From an evolutionary point of view our own group is composed of all our direct blood relatives, but in the real world human groups take many forms: family, friends, club members, neighbours, fellow villagers, the sports team, everyone who works for the same company – and so on and so forth, depending on what suits us best.

Sometimes this has sublime, noble consequences, such as when a whole group stands up in support of members who are subject to an external threat. One famous example in the case of the Netherlands comes from the early months of 1941, a bitterly cold winter in Nazi-occupied Amsterdam. The first round-ups of Amsterdam Jews led to what soon became known as the February Strike, the most significant act of protest in the Netherlands during the Second World War.

The story goes that shortly before the strike was broken, a slogan appeared on a wall somewhere in the city: 'Filthy Krauts, keep your filthy hands off our filthy Jews'. If the story is true, then whoever wrote that slogan was displaying, besides a sense of irony, an unusually profound insight into human relationships. It demonstrates how the distinction between members of the same species can either arouse protective instincts or provoke ruthless exclusion. Dutch Jews were and remained 'filthy Jews', but some filthy people were filthier than others. The filthy Krauts, for example.

This is where the existence of the uncanny valley returns to haunt us. It may have a useful effect in encouraging us to keep our distance from odd-looking and therefore possibly dangerous members of the same group, including clearly deranged, seriously ill or dead individuals, but it also creates in us a powerful aversion towards just about every aberration. Crooked noses, unfamiliar skin colours and narrow eyes – conspicuous characteristics have led Westerners to look askance at entire peoples. There is something not quite kosher about those who look that little bit different, or so our uncanny valley instincts warn us. We feel slightly uneasy in the company of people with features that are uncommon within our own group.

Exactly which features these are depends entirely on what's normal for the group in question. The Chinese find Westerners strange, Westerners are suspicious of black people, who in turn have their own ideas about the Chinese. Redheads stick out from the crowd almost everywhere and as children they can easily fall prey to ridicule and bullying. Dwarfism, bandy legs, baldness, hirsutism, conjoined eyebrows, albinism – you name it, any of these can cause suspicion, as can exotic clothing, strange eating habits or unfamiliar cultural etiquette.

Looked at from this perspective, racism and discrimination have a biological basis, however annoying that conclusion may be to the cultural relativists among us. They can comfort themselves with the thought that everyone is equally endowed with uncanny valley sensitivities, and our instinctive unease when confronted with people who are different from us need not have any serious consequences. If we so choose, we can simply brush our discomfort aside. But if the aberration in question is characteristic of a recognizable minority, if the circumstances are unfavourable, with significant conflicts of interest, and if people are systematically incited to hostility against a particular minority, then the results can be truly appalling.

The less noticeably others depart from the norm, the less impact the uncanny valley effect will have. We can't escape the strange smells coming from a new neighbour's kitchen – we simply have to get used to them – but no one is going to notice that the man who's just moved in next door is left-handed. This explains why left-handedness is generally regarded as only slightly negative, only a tiny bit suspicious. In folk tales the left is on average slightly more often connected with evil or with something frightening than the right, but it's certainly not inescapably associated with unadulterated evil. The same applies in everyday life. Many people pay lip service to the idea that left-handers are clumsy, but they don't really have anything against left-handed people. The subject of hand preference only faintly troubles them. Often they don't even know which of their friends and acquaintances are left-handed.

There are really only three areas in which we see what looks like a powerful rejection of everything connected with the left side and the left hand, and all three are strictly cultural and formal aspects of social life: etiquette, literature and religion.

To begin with religion, churches have strict regulations governing the hand with which ritual acts are to be performed and how the participants are to position themselves. In religious mythology a bad odour usually attaches to anything to do with the left. The Catholic Church has been known to interpret a refusal as an infant to suck a mother's left breast as evidence of piety in one of its saints. Catholic religious art is no less strictly rule-bound. A left-handed saint or apostle is inconceivable, let alone a left-handed member of the Holy Family. There are three reasons why these rules are so strict.

The first is the motif of inversion, which is used extensively by the Christian churches to accentuate the God–Devil antithesis. It makes the left literally devilish. Second, the great monotheistic religions refuse to have anything to do with magic and tend to distrust human passions, so they are hostile to the left side and the left hand. This is not to say that these religions are particularly geared to rationality – they aren't – but human impulses, being so hard to control, offer an ideal means of keeping fellow believers in check. Forbid people to succumb to natural urges and the faithful are bound to fail. The celibacy of Catholic priests and the separation of the sexes advocated by Islam are the most extreme examples, but dietary laws, compulsory periods of fasting and an endless number of other behavioural rules, preferably as complicated and as inimical to relaxation, fun and enjoyment as possible, are part of the cherished arsenal of practically all religions. The sense of guilt that flows from the inevitable violation of these rules helps faiths to discipline their flocks and thereby to perpetuate themselves.

The third reason, the formality of church rituals and religious art, is more abstract. Rituals are unfathomable almost by definition. If they are not to become garbled or forgotten, they must be clearly recognizable and repeatable. Rituals have less to do with the actual acts involved than with the fact that those acts have to be performed in precisely the prescribed manner. Anyone who deviates from the course laid down automatically invalidates the ritual. It's logical that in the world of rituals, contrasts have to be made as black-and-white as possible, and as a result the left side and the left hand are not just a tiny bit suspect but subject to strict prohibitions.

Icons and other religious depictions are also in some sense ritualistic in nature. They need to convey a clearly defined religious message that can be understood in the absence of any additional information. Clarity is paramount. A vast gulf therefore exists between the artist and the public; there's little or no room for anything personal. Only in the West has this changed to some degree, under the influence of the Renaissance. In most religious art the viewer must be able to deduce precisely who is being depicted and what the image means, based purely on the symbols contained within the work.

Exactly the same rules apply to etiquette, where the right hand is the 'good hand', the guest of honour sits at table to the right of the host or hostess and so on. The less intimately people know each other, the easier it is for misunderstandings to arise, so it's important that when relative strangers meet no one acts in a way that's unexpected or hard to interpret, unless they have good reasons for doing so. We act formally and adhere to set rules. In theory it doesn't matter what those rules are, as long as they can be clearly recognized and easily remembered. Black-and-white, therefore, is the message. The natural preponderance of right-handers automatically makes left-handedness anathema.

In literature, finally, whatever the language, left often means clumsy, duplicitous and sly. Presumably this flows directly from the fact that for centuries literature reflected the established opinions of church and state, so that as a whole it deployed symbolism approved from on high. In fact it did a fair amount to canonize a culture's officially sanctioned symbolism and set it in stone. Literature has always been an important source for lexicographers and teachers, and through them the negative connotations of a language are perpetuated and reinforced. All things considered, it's no wonder that formalized metaphors involving the left as a concept are a good deal harsher and more negative than can be justified by anything in our everyday experience.

9
Witchcraft and Pogroms

Devilish practices and black magic clearly belong to the realm of inversion. At a witches' sabbath everything moves anticlockwise, in a direct reversal of the customs of the church. The Devil, damned rascal that he is, demands to have his arse kissed by his followers. Unholy black replaces virgin white and the congregation worships evil instead of good, death instead of life, lechery instead of fidelity and abstinence. Black magic has no characteristics of its own; it consists purely of negation and is therefore derivative, taking its form from the precise reversal of what is regarded as valuable and holy in any normal religion. So just as it seems natural that the bad guy in an old-fashioned cowboy film wears a black hat, in European art the Devil and his accomplices are left-handed. In the modern world this has little significance, but it may have made an unfortunate impression in the past. Left-handed people might quite easily have been seen as confederates of the Devil in panicky times, for example when people were obsessed by a fear of witches, and ordinary citizens were accused, condemned and killed on even more tenuous grounds.

Yet astonishingly this never happened. A group of comparable size, the homosexual segment of the population, was persecuted with grim determination down through the centuries. Jews, gypsies and tramps have all been targeted from time to time, but the left-handed were spared persecution. Even in those thousands of witch-trials, a defendant's left-handedness was hardly ever presented as proof, or even as corroborating evidence of contact with the evil one, unlike any number of other features found on the left side of the body, such as birthmarks, warts and scars, which were far less likely to be discovered.

The main thing that saved left-handers was probably the fact that they never presented themselves as a group. Jews and gypsies grouped

A witch kisses
the Devil in the
appropriate place.

up quite noticeably, separating themselves from the rest of society in various ways that were impossible to ignore, sensing that they were a distinct people and presenting themselves as such, with all the advantages and disadvantages that entailed. It's always dangerous to stand out. Adopting an isolated position makes the reserve that naturally smoulders among the general population as a result of the uncanny valley effect flare up into a dangerous form of jealousy.

Groups that wilfully separate themselves arouse curiosity. The distrustful vulture instinct makes everyone else wonder: 'What are they up to?' People feel shut out and think to themselves: they're acting so secretively, they must be in possession of something interesting that they're trying to keep for themselves. It's a mechanism seen in operation daily in every school playground. If it doesn't quickly become clear what a group has that's so desirable – and it rarely becomes clear, because there's generally nothing involved that could appeal to non-members – then the most dire suspicions arise. People begin to think it's less a matter of a pact formed to protect and guard a desirable possession and more a conspiracy against the rest of society. Once this stage is reached, the jealousy and frustration of those who feel excluded become combined with fear – an explosive mix. It's at this point that strange, frightening stories start to circulate, and to be believed: Freemasons sacrifice babies at their meetings; Jews dine on Christian children;

gypsies will steal your offspring and sell them as chimney sweeps or worse. Those people must worship the Devil, otherwise why would they be so secretive?

Odd or exceptional groups are liable to be made scapegoats for events that have nothing to do with them. You can easily blame a bunch of people who are strange and secretive for a failed harvest or some such natural disaster. The conspiracy of Jewish financiers planning to take over the world, which the Nazis believed they had uncovered, was nothing more, and nothing less, than a modern instalment of a serial drama that had been running for centuries, each episode leading sooner or later to the same outcome: violent death, where possible on a grand scale.

The left-handed don't feel themselves to be a group in any sense. Even in this era of effortless global communication, it's proven impossible to create a viable association of left-handed people in Europe. Only in America does anything of the kind exist. Just as right-handers will often fail to notice that an acquaintance is left-handed, it may take two left-handers a long time to discover they have a hand preference in common. It's probably this lack of solidarity, the absence of any group feeling or group behaviour, that explains why the left-handed have never attracted the kind of attention that's proven fatal to so many other groups.

A second reason probably lies in the size of the category. Around 10 per cent of people are left-handed and they are fairly evenly spread across the population. In some families the percentage is rather higher, in others lower, and the phenomenon occurs slightly more often in men than in women – a difference that's measurable but too small for anyone to notice in daily life. There the variability ends. Everyone's circle of family, friends and acquaintances includes one or more left-handers, so persecution would come at considerable cost: sooner or later it would inevitably mean personal losses for us all. That's precisely the opposite of what pogroms and so forth are all about: us against them with the two kept as strictly separate as possible. That could never happen with attacks on left-handers. There are simply too many of them, and they're too evenly spread.

So what about homosexuals? They too make up roughly 10 per cent of the population and there's no greater concentration of them in any particular group, yet they are one of the most persecuted human categories in history, even down to the present day. One obvious reason for the difference in treatment is that homosexuality is no mere symbolic violation of the natural order of things but rather a matter of actual behaviour, even if it's less what homosexuals do with each other than the thought of it that makes the heterosexual majority so nervous.

We respond as we do because our personal and social lives, from our individual identities to the security provided by marriage, and by familial rights and duties, depend entirely upon ordered sexual relations. Any breach causes chaos, which we loathe as cats loathe cold water. Adultery threatens the permanency of relationships and therefore inadvertently endangers the integrity of a family's property. Divorce actually erodes that property, which is why in many parts of the world it's regarded as worse than adultery. Instead of getting divorced, society would prefer you to keep quiet about your bit on the side, even in the case of a long-term extra-marital relationship. Two things demonstrate that this is very much bound up with money and with the security of existence. First, until Napoleonic times marriage was mainly intended for the propertied classes; those who lived hand to mouth simply moved in together without a marriage certificate. Second, in our modern world there's a clear link between prosperity and conjugal ethics. The richer the society and the more security it offers, the likelier people are to experience a bond with a partner as galling and terminate it.

Worse even than adultery and divorce are the fruits of illegitimate relationships. Bastard children tangle the web of kinship. We can no longer be certain who is truly 'one of us' or even exactly who 'we' are. The effects of incest are even more disastrous, especially when children are born as a result. Incest is fiercely condemned more for this reason than because of fears for the health of the offspring. Although not free of risk, semi-incestuous relationships, between cousins for example, are normal in many cultures as long as they are sanctioned by marriage.

Homosexuality brings us even closer to complete disarray. It not only crosses the boundaries of the sexual order, always a precarious structure, it cocks a snook at it, tossing everything upside-down, even the foundation on which it all rests: the formation of pairs between the two sexes. Homosexuality inverts the most central of all sexual values, in a 'devilish' reversal that seems to strike at the root of the continued existence of the species. It's as if homosexuals provide the rest of humanity with a glimpse into the miasma beneath our feet. They are not thanked for it.

From a rational point of view, all this nervousness is sheer nonsense. The human species isn't going to be wiped out by a bit of gay sex, but this obvious fact does nothing to reduce the fear, which is the source of some shocking aggression, especially from the most violent and volatile of citizens: young men who feel uncertain about their sexuality and their place in the world. The more anxious a society becomes about sex, and the more problematic and emotionally charged direct contact between men and women is as a result, the louder its young men will shout about honour and pride. They'll be all the more frightened of homosexuality and hit out all the harder against it.

One thing that makes the position of gay people a good deal more difficult still is that homosexuality can be concealed. Keeping left-handedness secret is almost impossible. The hiding of their proclivity only makes homosexuals as a group more vulnerable to persecution, since it means their numbers can be grossly underestimated. With a few exceptions, this has happened all over the world in all eras. It's meant that heterosexuals could bask in the illusion that gay sex did not happen in their own circles, and therefore that gay people could safely be persecuted. It goes without saying that this quite often led to unpleasant surprises, but they too were hushed up wherever possible. So great is the taboo that even today in large parts of the world it's easy to find people who'll swear, hand on heart, that homosexuality doesn't exist in their country. In such circumstances it's fairly easy to make it a capital offence.

Anyone who thinks witchcraft and witch-hunts are a thing of the past is sadly mistaken, and I don't mean the fashionable tinkering with white magic that's become popular in Western Europe among ladies with time on their hands and esoteric interests. I mean the real, old-fashioned stuff. Mind you, what is real? Not the witches. They tended to be silly innocents with the misfortune to have attracted the wrong kind of attention for one reason or another. The persecution, though, was and still is a bitter reality.

In and around Kenya and Tanzania, the victims are mainly albino children. They are seen as possessed by evil powers and subjected to murderous violence as a result. Nastier still is the lucrative trade in albino-based magical remedies that emerged in the late twentieth century. Albino children are slaughtered for their body parts, which are then boiled to make soup.

In Nigeria too, children are systematically branded witches and excluded, mistreated or even killed. Their inherent childish innocence makes them an attractive target, since it's all the more shocking and frightening when an innocent creature turns out to harbour evil. The witch-hunters are spiritual leaders, usually with excessively devout

Christian backgrounds, who feel a need to create a distinctive profile for themselves, to use a high-sounding Western term. Child witch-hunts are a distorted form of aggressive evangelism and the persecution begins not so much because the appearance or behaviour of the victim has attracted attention as because the zealot has something to gain. The invention of an intangible threat that might pop up anywhere without warning is a tried and tested means of keeping the flock together and making it toe the line, especially if you can then emerge as its saviour, the only person who, as a special agent of the Lord, can recognize and neutralize the danger. Meanwhile, along with the respect and gratitude of believers, a great deal of money will come your way, since driving out evil is not something you can be expected to do on the cheap. So the exorcist's knife cuts both ways.

Elsewhere in the world it's always been mainly women who fall prey to witch-hunts, often widows who have a rather marginal position in society and lives of poverty and solitude. In India hundreds of people die every year after being branded as witches by their neighbours. In that big, crowded country no one paid much attention until the nation was suddenly caught with its trousers down in the autumn of 2009. Five women in a village called Pattharghatia were chased through the streets naked, whipped and made to eat excrement by a frenzied crowd several hundred strong. All this took place at the instigation of six other women, who claimed they had the gift of the holy spirit and were therefore able to recognize the witches who were about to bring calamity upon the village. For once the police stepped in and arrested eleven villagers, including the six who had claimed supernatural powers yet astonishingly were not witches themselves. The reason for the intervention of the authorities was that someone had made a video recording of the event, which was shown nationwide on television.

There are countless possible grounds for concluding that someone is a witch. A frightening appearance – albinism being one example – or a marginal position in society are often regarded as signs, but any deviation from the norm will do, anything that might get a person noticed. There is one aberration that's never involved, as far as we know: left-handedness. It wasn't even a factor in the witch trials that wreaked havoc in Europe several centuries ago. This isn't so odd as it may at first appear, since the widespread persecution of alleged accomplices of the Devil, his female accomplices in particular, did not arise out of the blue.

The persecution of witches began in the fifteenth century, when the old medieval structures were finally starting to crumble and art and science were wriggling out from under the suffocating constraints of the Christian authorities, a period known as the Renaissance. Furthermore, the urban mercantile classes were demanding a place on the political stage. Their power lay in their money, rather than in any medieval appeal for loyalty to a system ordained by God, and they were determined to win. Eventually, centuries later, that shift in power would result in two revolutions: the Industrial Revolution that made industrial production and capital the most important forces in creating and sustaining the powerful of the earth, and the French Revolution, which finally put paid to the invulnerability of a mythical authority that had been placed above everyone by God and tradition. These changes were still in the distant future. A fifteenth-century European lived in a world that was beginning to rock on its foundations, a world that still bore the barely healed scars of the universal upheaval brought about by the Black Death of 1347–51. That pandemic – traditionally understood to have been bubonic plague although it may equally well have been Ebola or possibly the Marburg virus – claimed 75 million, perhaps 100 million lives on the continent in the space of a few years. The loss would not be made up until around 1600 and the whole of Europe was left deeply traumatized. As if that were not enough, roughly simultaneously a period of relatively low temperatures began, known as the Little Ice Age, that lasted until the nineteenth century and caused harvests in large swathes of Europe to fail year after year, resulting in famine, price explosions and further epidemics. In this climate of hunger, want and social uncertainty, a shaken population felt a need to find scapegoats. A deep mistrust of minorities arose, fuelled by a fear of evils that apparently struck without rhyme or reason. By about 1430 a world had been created in which people were quick to see a Devil's disciple around every corner, a member of a diabolical worldwide conspiracy – a witch.

It was also a period in which little remained of the authority of the church over worldly affairs. This was largely the fault of the church itself. Long years of infighting and political intrigue had caused deep cracks to appear in the image of papal inviolability. Antipopes, corruption and libertines within the church who were vulnerable to blackmail gradually led most worldly rulers to pay a good deal less attention to what the pope said, which is not to deny that those rulers were often deeply religious, either by personal conviction or for reasons of political opportunism. Nevertheless it was 1534 before Henry VIII openly and irrevocably defied the pope and the church hierarchy, setting up a religious enterprise of his own as boss of the Anglican church.

By this time something had been brewing among ordinary believers for many years. The signs had been visible for two centuries: movements of dissenters such as the Waldenese and the Cathars came and went, as did internal reformists, the Cluny movement being one example. Dissatisfaction with the ethical state of affairs within the church led to the establishment of various new mendicant orders, including the Franciscans. In the end none of this helped. The church remained corrupt; indeed if anything it became even more degenerate than before. As 1500 approached the tipping point arrived: the Reformation was now on its way and it would ultimately lead to the eighteenth-century creation of the nation state as we know it, ending the role of religion as the most important guarantee of political unity.

In the desperately unstable situation that resulted from this power vacuum, with everyone searching for a new equilibrium and new values, the ruling powers did all they could to survive and maintain their grip on power. They naturally turned to the oldest, indeed the only available mechanism: religion. Under its banner the battle lasted for many years, with bloody religious wars ravaging Europe from 1517 to 1648. This was the heyday of the Inquisition. And of the obsession with witches.

Fear of witches, which had always existed but from about 1430 spilled over to become a general fear of a worldwide conspiracy of diabolical powers, was whipped up mainly by those at the top of the hierarchy. The persecution of witches developed into a veritable holy crusade, whose background was more political than emotional. It was one of the means by which the authorities, consciously or unconsciously, tried to prevent the collapse of the social order. As yet they had no other binding agent to hand. Where the new concept of the sovereign state did at last catch on, witch-hunts quickly ceased. This happened first in the Republic of the United Netherlands, which was founded in the late sixteenth century out of Spanish possessions along the North Sea coast. In 1603 in Nijmegen the last death sentence was carried out, whereas in the nearby province of Limburg, which did not belong to the young Republic, 70 witches were executed between 1603 and 1637. In Spain and Italy, where the Reformation failed to gain a firm footing and the old faith remained fairly securely in place, the persecutions ended in about 1620 and relatively few death sentences were passed. Elsewhere conditions were much more chaotic, and witch-hunts were more ferocious and went on until a much later date. In Western Europe they lasted until not long after the Peace of Münster in 1648, which brought an end to the wars of religion, and they continued longest of all in

Eastern Europe, where witch trials were rife until well into the eighteenth century.

Further evidence that the obsession with witches was not simply a case of mass hysteria but was conscientiously exploited, kept on the boil and sometimes even orchestrated by the religious and secular authorities is the almost endless series of handbooks about the witch problem that appeared in this period. The most notorious example is *The Hammer of Witches*, or *Malleus Maleficarum* to give it its original Latin title, published in 1486 and a runaway success that was reprinted dozens of times. The book is controversial in every respect and even its authorship is uncertain. Two Dominican inquisitors were credited with writing it, although they were not named on the title page until after both of them had died. One was the fanatical persecutor of witches and heretics Heinrich Kramer, a man with a slightly dubious reputation and few scruples, who preferred, as was usual among scholars of his time, to go by a Latin version of his name, Henricus Institoris. As a somewhat scatterbrained fellow, an unguided missile, not averse to the occasional sophism, he seems a rather strange partner to his fellow author Jacob Sprenger. Although Sprenger's attitude was barely any more enlightened than Kramer's, he was regarded as a respectable and sensible man, and as an inquisitor he had far greater authority. It has

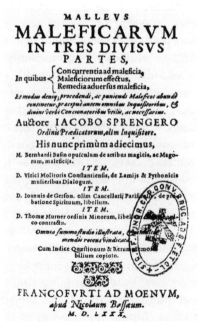

Title page in an edition of *The Hammer of Witches* (1580), published in Frankfurt am Main, on which Jacob Sprenger is the only author mentioned. In reality Heinrich Kramer was the main if not the sole author.

been established beyond doubt that unlike Kramer he was never involved in a witch trial that ended in execution.

Since in many ways *The Hammer of Witches* is a bizarre treatise that frequently contradicts itself, and because it was generally known that Sprenger refused to have anything to do with Kramer, it has been claimed that Sprenger did not in reality have a hand in the writing of the book. Kramer is said to have entered his name as an author to give his tome greater cachet. This is improbable, however. *The Hammer of Witches* was a huge success from the moment of its publication, going through reprint after reprint and becoming greatly respected in witch-hunting circles, which included many German princes. Everyone who was anyone knew it had been written by Kramer and Sprenger, and Sprenger never uttered a word about Kramer's supposed misattribution. Of course it's possible that Sprenger did not contribute to the actual writing of it, but in that case he must have been involved in some other way. At the very least he consciously tolerated Kramer's use of his name.

The content of the book is no less odd than the mysteries surrounding its authorship. To call it a bizarre piece of work is almost an understatement. It is utter claptrap from cover to cover, a dizzying labyrinth of fantasies about incubuses who inseminate fallen women with evil, witches who tear off men's penises and stick them back on in a trice, and plenty more charming nonsense of that sort. Yet for all its deadly earnest craziness it does offer a precise definition of what witches actually are and describes in great detail the process that must be adhered to in trying them. Surprisingly, perhaps, the court proceedings, if we overlook a few sloppy and self-contradictory passages, are relatively well thought through and hemmed in by all kinds of safeguards. Torture is permissible, but it's certainly not presented as a first resort for those attempting to get at the truth. Evidence derived exclusively from torture is regarded as inadequate. Grounds for prosecution do not include, for example, a vague alliance with the Devil. Only the most serious form of apostasy counts: a true pact with Satan, sealed by coitus. We are not told how such a thing could be proven unless the woman confessed. Crucially, there had to be an effective *maleficium*, in other words actual harm done to specific people. To put it in more modern terms, reasonable suspicion of a punishable offence was required.

Physical marks on a woman's body were hardly ever presented as proof in formal witch trials. They served purely as corroborating evidence for the prosecution's case. Ordinary people as well as informers and self-appointed witch hunters had a different attitude, however. Unusual warts had a considerable 'what did I tell you?' impact, as did strangely shaped scars, peculiar birthmarks and any number of deformities great and small. Rudimentary extra nipples had a particularly powerful impact. These are fully or partially developed supernumerary nipples on the 'milk lines', the two parallel lines of tissue running down across the chest and stomach, from which nipples sprout in all mammals, including humans. The chances of coming upon such nipples are quite high, since they can be found in about one person in a hundred, sometimes purely as a discolouration, sometimes fully functional. They were suspect not only because of their association with intimate motherly care and eroticism but because by their very nature they normally remain hidden under clothing. In a climate of suspicion it was a short step to the conclusion that the owner must be hiding them deliberately, for nefarious reasons.

We should not be too surprised that left-handedness as such never featured even as corroborating evidence. There's nothing particularly secretive about it. It may well pass unnoticed for years, but at the same time it is hard to conceal and will not suddenly emerge to the utter astonishment of others in the way that an extra nipple or a suspiciously shaped birthmark on the buttock might. In any case, it's simply too common. Had left-handers been persecuted, all citizens would have run an unacceptable risk of losing someone close to them.

10
Factionalism

'In such manner labour the National Deputies . . . with toil and noise,' wrote English satirist and historian Thomas Carlyle in 1837 of the revolutionary French parliament of 1789, 'cutting asunder ancient intolerable bonds; and, for new ones, assiduously spinning ropes of sand. Were their labours a nothing or a something, yet the eyes of all France being reverently fixed on them, History can never very long leave them altogether out of sight.' The Assemblée is a mess, says Carlyle:

> As many as a hundred members are on their feet at once; no rule in making motions, or only commencements of a rule; Spectators' Gallery allowed to applaud, and even to hiss; President, appointed once a fortnight, raising many times no serene head above the waves.

There is hope all the same. The chaos is not unlimited:

> Nevertheless, as in all human Assemblages, like does begin arranging itself to like; the perennial rule, *Ubi homines sunt modi sunt*, proves valid. Rudiments of Methods disclose themselves; rudiments of Parties. There is a Right Side (Cote Droit), a Left Side (Cote Gauche); sitting on M. le President's right hand, or on his left: the Cote Droit conservative; the Cote Gauche destructive.

The rightist faction was not composed of sheep, by any means:

> On the Right Side, pleads and perorates Cazales, the Dragoon-captain, eloquent, mildly fervent; earning for himself the shadow

of a name. There also blusters Barrel-Mirabeau, the Younger Mirabeau, not without wit: dusky d'Espremenil does nothing but sniff and ejaculate; might, it is fondly thought, lay prostrate the Elder Mirabeau himself, would he but try, which he does not. Last and greatest, see, for one moment, the Abbé Maury; with his jesuitic eyes, his impassive brass face, 'image of all the cardinal sins'. Indomitable, unquenchable, he fights jesuitico-rhetorically; with toughest lungs and heart; for Throne, especially for Altar and Tithes. So that a shrill voice exclaims once, from the Gallery: 'Messieurs of the Clergy, you have to be shaved; if you wriggle too much, you will get cut.'

Carlyle spotted an ominous figure on the left:

seagreen Robespierre; throwing in his light weight, with decision, not yet with effect. A thin lean Puritan and Precisian; he would make away with formulas; yet lives, moves, and has his being, wholly in formulas, of another sort. 'Peuple,' such according to Robespierre ought to be the Royal method of promulgating laws, 'Peuple, this is the Law I have framed for thee; dost thou accept it?', answered from Right Side, from Centre and Left, by inextinguishable laughter. Yet men of insight discern that the Seagreen may by chance go far: 'this man,' observes Mirabeau, 'will do somewhat; he believes every word he says.'

In this warm-blooded description of the birth pangs of a new and revolutionary France lie the roots of the left–right dichotomy that has come to dominate modern society like no other. The political, ideological left and right have ever since been the benchmark of political and social polarization from London to Beijing, from Puget Sound to Patagonia. The French delegates invented the concepts more or less by accident, but it was Carlyle who chiselled them into granite.

Carlyle characterized the leftist faction as destructive. He meant it was the place for people who wanted to change things, to build a new, better society, and who therefore demanded that the old structures of authority be torn down – precisely those structures 'the right' held dear. Right-wingers thought society was fine the way it was, with altars and tithes and respect for one's betters.

In the Paris of 1789 that was the basic antithesis, but people would not have been people if they hadn't immediately started to deck out the core concepts of left and right with countless associations, encouraged by events and by the emergence of a wide range of tendencies and ideologies. Little by little a new table of opposites was created, looking something like this:

LEFT	RIGHT
pro-change	conservative
egalitarian	authoritarian
group-oriented	individualistic
disobedient	law-abiding
experimental	conformist
pacifist	militaristic
poor	rich
supportive	self-seeking
rational	myth-oriented
politically engaged	politically aloof
permissive	intolerant
libertine	libertarian
dynamic	static
open	closed

This is no less inconsistent, generalizing, simplistic and contradictory than Pythagoras' list, or any other version – and no better a reflection of reality – yet it does offer a clear image of the lines along which Western political and social discourse has run for the past two centuries.

In one fascinating respect it's fundamentally different from Pythagoras' table, which was concerned with far more neutral phenomena, most of them aspects of nature. Straight, crooked, sun, moon, life, death – these things may be bound up with emotions and value judgements but in themselves they are objective concepts, independent of mankind. The table of socio-political opposites contains quite different elements. Almost without exception they are concepts invented by man that could not exist without him. These are choices made for reasons of principle, indissolubly linked to value judgements, yet we use both tables opportunistically, simply shutting our eyes to their contradictions and inconsistencies.

Take respect for authority, traditionally seen as one of the core values of the conservative right. Since the rise of socialism (left) and capitalism (right), if not before, the right has been associated with individualism, in the sense that it's oriented towards self-interest rather than the general interest. But how does that fit with the equally right-wing tendency towards conformism? What does individualism mean if at the same time you always have to conform? Conversely, how can leftists be at one and the same time socially oriented, and therefore the personification of solidarity, and non-conformist? Which societies were the most egalitarian in the past? That particular honour goes to the Communist regimes in Russia and China and their satellites. And what should we make of Nazi Germany? It was the textbook example of an extreme right-wing society but it called itself socialist, was explicitly anti-capitalist and indeed had all the familiar features of socialism.

Let's be honest. Countries that are so intolerant that you'd do well to avoid getting noticed, where conformism is therefore a prerequisite for survival, have always ranged right across the spectrum from the extreme left to the extreme right, no matter how those terms are understood. The interesting thing is that although a table of opposites like this has little to do with reality even when it comes to the basic principles according to which we live, it is felt to be a helpful way of organizing our thoughts in this complex chaos we call society, which explains why we're so quick to use the terms left and right, and why we often find it so incredibly hard to say exactly what we mean by them.

In modern, more or less democratic countries with a high degree of press freedom, the concept of the political left has been used of late in one particular way. Complaints are consistently heard that the media in these countries are too left-wing. Strangely enough you rarely if ever hear claims that the press or television are generally speaking too far to the right.

This tends to contradict the impression we get from reading the better British newspapers, for example, and it's hard to discern any features of the BBC that would identify it as the mouthpiece of socialist agitators. In American dailies you'll have to look long and hard to detect voices that sound the slightest bit leftist, even in the best-quality and most critical papers. The biggest selling Dutch newspaper, *De Telegraaf*, calls itself 'the paper of the wide-awake Netherlands'. The term 'wide-awake' refers to the early to bed, early to rise, hardworking conformist citizen, and the paper has an undisguised right-wing slant. Things are little different in France. Wherever you look, the lower reaches of the media, especially the most popular British newspapers, tend overwhelmingly towards the narrow-minded, fearful right and show very little interest in improving society. Instead of dealing with troublesome social evils they concentrate on bite-sized personal scandals and simple sensation.

What 'too left-wing' means is that newspapers and news or current affairs programmes on radio and television often report on, and side with, those who are victims in one way or another. They are full of stories about the powerless and neglected, the oppressed little man and those who are dependent on others. They report on the weak, the sick and the homeless, refugees and illegals, butterflies and beetles. This is indeed how the media work, but ultimately it has little to do with left and right. Good-news media and nothing-to-worry-about media don't survive for long. Whatever their complexion, media organizations rely on bad news, scandals, problems and abuses, since they make for interesting reports and revelations. These are the things people want to read about.

The result is that newspapers and magazines, radio and television, plus whatever other media we have or are currently inventing, always pay disproportionate attention to the things that disturb and frighten us, to the bad things in life. Those are precisely the things that the conservative, contented segment of the nation would prefer not to think about too much, which is something they have in common with the people who hold high office at any given time and are therefore seen as responsible for everything that's wrong with society. They, the comfortable and the powerful, whatever their other beliefs or political persuasions, are the source of complaints that the media are too far to the left.

This eternal criticism of the media therefore has little to do with left or right. American linguist and political thinker Noam Chomsky has uncovered its true purpose. It always suits the class of conservative top dogs in any society to say that the media are too left-wing, because by doing so they define what counts as reasonable opposition. If the respectable media are too far to the left, then truly critical or negative opinions about the political and social state of affairs automatically belong to what the British rather delightfully call the loony left.

This is a fairly easy way to declare off-limits any critical voices calling for radical change to the existing social order. After all, everyone is aware of the boundaries, everyone follows the mainstream media to some extent. No one wants to be sidelined, so we all, especially those of us who are involved in politics, automatically stay on the safe middle ground, where uncomfortable matters can be swallowed up in a sea of obfuscation. Those who complain about media bias today may be more civilized and subtle in their approach, but as an instrument of power for use by the ruling class they perform the same role as the Stasi did in the former DDR. Their goal is to set limits to critical thought and keep control of movements favouring change.

11

The Ideal Warrior

The Benjamite story about the 700 left-handed slingmen demonstrates that left-handedness was a serious problem in warfare even thousands of years ago. Why else would anyone take the trouble to gather so many left-handed people together and organize them into a military unit of their own? The problem still exists today, in fact it has barely changed.

Organized military action demands a great deal of the soldier, and left-handedness is undesirable. No matter how inconspicuously left-handers tend to function in normal circumstances, in battle they suddenly become a real danger to their comrades. A group of hooligans can turn itself into an army only through discipline, by doing exactly the same thing on command, like a machine. This kind of discipline gives it so much power and effectiveness that it can defeat poorly organized opponents, even those that are theoretically far stronger. The ancient Greeks knew this, with their phalanxes of lancers who did little else but dourly persist in walking side by side with their long lances thrust out ahead of them. That orderly block of men, bristling with iron spikes, simply steamrollered across everything that stood in its way. Using roughly the same approach, but with short swords sticking through their tightly linked walls of rectangular shields like the flails of a gigantic threshing machine, Roman legions hacked their bloody path through half the known world. When serious trouble erupts, these tactics are still used today by units of riot police all over the world, except that they wield batons rather than the deadly Roman *gladius*. On all essential points the approach is the same, with serried ranks of anonymous men steered as a single unit like automatons. Military combat worked this way right up until the First World War, when the machine gun put an end to a military tradition that was almost 3,000 years old.

Up to that point, the left-handed soldier presented a serious problem. Effective armies depended on two things: discipline, which ensured that when push came to shove soldiers would not run away, and uniformity. Left-handedness violated that uniformity.

Before the invention of gunpowder, a disciplined unit of infantry depended on the relatively effective protection of an unbroken line of shields. Imagine a left-handed soldier coming to stand among them, holding his shield in his right hand. Not only would the left-hander create a dangerous gap in the unit's cover, he would seriously obstruct the man to his right. That's not all. A right-handed soldier moves to the left as he steps forward, so that he can strike from behind his shield, but in that position left-handed soldiers are at their most vulnerable. They will instinctively move in the other direction, causing chaos among the men. On commands such as 'right turn' and 'left turn' a soldier's equipment can all too easily get tangled up with another's. In short, with an obdurate left-hander in the ranks, who needs enemies?

Even today the military world continually runs up against this problem. The bolt of a rifle is on the wrong side for left-handers, so it takes them longer to reload. All standard rifles eject empty shells to the right, away from a right-handed shooter. A left-hander holds the gun to the left shoulder and so runs the risk of hot cartridge cases flying at his or her face, or right arm, which supports the barrel. Some modern firearms, though not all, are designed in such a way as to avoid this problem, but even so the left-handed soldier is at a disadvantage because the butt is shaped to fit the right shoulder. The left shoulder against which he or she braces a rifle therefore has to put up with more punishment than strictly necessary. Left-handed models are not the solution. When called upon to do so, soldiers have to be able to use the gun carried by the soldier next to them. There's no time to work out how best to handle an adapted rifle; you must grab it and use it without having to think. So even in the most modern of wars, uniformity, consistency and predictability are essential to any successful fighting unit. A left-handed soldier does not fit in.

There are some circumstances in which a left-handed soldier may be useful. As a right-hander you can shoot around the end of a wall to your left, but if the wall lies to your right you are forced to expose yourself to the enemy's line of fire. Here left-handers are in their element. There have always been people in army circles who dreamed of an ambidextrous soldier, the ideal warrior, who could wield a sword, pike or shield with the left hand as easily as with the right, mount a horse from either side with equal ease, and shoot just as effortlessly and accurately from either shoulder.

In the final decades of the nineteenth century it seemed as if that dream was about to become a reality. It was a time of great optimism and faith in progress. On the political front Europe was relatively peaceful between 1871 and 1914, partly as a result of a complex system of treaties and alliances. True, in Russia and the Far East there were conflicts of various kinds, and from 1899 to 1902 the Boer War raged in South Africa, but that all seemed little more than a bit of bother at the world's margins. The West was proud of its unprecedented level of civilization, not yet defiled by the two disastrous world wars of the twentieth century. Railways had opened up the possibility of long-distance overland travel and immense construction projects like the Suez Canal and the Panama Canal had proven how far man could bend nature to his will. In less than a century the steamship and the steam train had made the whole world accessible. The first cars were starting to crawl around, and in 1903 mankind took off when Wilbur and Orville Wright made their first powered flights on the beach at Kitty Hawk. There seemed to be no limit to what technology could do. The world was whatever we made of it.

At the same time impressive advances were being achieved in science: so impressive that Lord Kelvin, discoverer of absolute zero and the driving force behind the laying of the first transatlantic telegraph cable, was convinced that physics had virtually completed its task. 'There is nothing new to be discovered in physics now. All that remains is more and more precise measurement', he prophesied. He had no inkling, convinced as he was that 'x-ray is a hoax', of the discoveries that were just round the corner on the subatomic level. In retrospect the game proper had yet to begin.

Even by this point Charles Darwin's evolutionary theory was starting to have far-reaching consequences. Although his theory was misunderstood more often than not, two things had penetrated the consciousness of the general public: human beings were closely related to animals, especially apes, and people were not endowed by God with eternal and immutable characteristics but had evolved gradually over time. They could therefore change. Why should they not be malleable, at least to some extent?

Under the influence of all of this, a new kind of interest developed in the phenomenon of right- and left-handedness. If human characteristics were not set in stone, then surely it must be possible to change them at will. Our closest relatives, the apes, showed no sign of hand preference, so why would we not be capable of creating, through training, a two-handed, more complete person? The fact that apes had not developed to anything like the same extent as humans was simply ignored.

At the start of the twentieth century this attitude led to the founding of an association, the Ambidextral Culture Society, which gained a considerable following in fashionable circles. Devotees of two-handedness were convinced that one-handedness was the last serious handicap preventing us from achieving the supreme ideal of the modern, highly educated and well brought-up person, representing complete self-fulfilment. Children, as long as they were correctly trained in the equal use of their two hands, would grow into perfect people, no longer hampered by a preference either way. It remained unclear why hand preference was such a dreadful thing, aside from the fact that the human being apparently has two virtually identical hands. In this sense the Ambidextral Culture Society was a worthy forerunner of later all-embracing fads such as the fashion for anti-authoritarian childrearing that prevailed in the early 1970s, or the Bhagwan movement of the subsequent decade.

As is characteristic of movements of this sort, the ambidextrous ideal lacked a well-considered foundation. People rushed to claim prominent figures from history as its posthumous defenders. The great eighteenth-century French philosopher Jean-Jacques Rousseau featured yet again among the chosen few. A myth was born. Even today, long after the ambidextrous movement faded into history, he is still regularly presented as an advocate of a two-handed upbringing.

Rousseau's supposed predilection for ambidextrousness is based on a passage in his 1762 book *Emile* in which he writes about the development and raising of children. He says that the only habit a child should learn is the habit of not having habits. He should not be carried consistently on one arm or the other, he should be encouraged to shake either hand, and in general he should be allowed to use both hands without distinction. Poor Rousseau. He was simply voicing a protest against the conventional pressure to use the 'correct hand' and appealing for children to be allowed to choose for themselves, but his readers interpreted his approach as grounds for forcing children to be ambidextrous. Rousseau's appeal for permissiveness was effortlessly translated into a demand for coercion.

The driving force behind the movement was a certain John Jackson, who in 1905 published a passionate appeal for two-handedness and a two-handed upbringing under the title *Ambidexterity*. Its foreword was written by no less prominent a figure than Lieutenant General Lord Baden-Powell, founder of the scouting movement and hero of the wars against the Ashanti in the closing years of the nineteenth century – complete with a double signature, written simultaneously with both hands.

Baden-Powell was a man with far-reaching ideas about many things. He had a strong faith in the optimal training of the human body, which for a military man was of course only sensible. On the one hand this gave rise to the scouting movement as a means of developing the average boy's body to perfection. On the other, he was convinced that one-sidedness in people, whether it was a matter of hands, legs or eyes, was a serious obstacle to the attainment of the perfection he demanded in a soldier. As far as Baden-Powell was concerned the importance of two-handed training from an early age was almost impossible to overstate. He claimed he could keep on top of his office work only by regularly switching his writing hand and he regretted that as a child he had not practiced writing about two different subjects at once. If for that reason alone, he cannot have been much of an intellectual. Even English lords have only one head and he doesn't explain how his could have busied itself with two subjects at a time, each independently of the other.

Baden-Powell's beliefs about ambidextrousness left their mark in at least one scouting custom: scouts greet each other with the left hand. This habit is only tangentially connected with hand preference, incidentally. The idea occurred to the child-loving warhorse in 1896, when a defeated Ashanti chief reverentially offered him his left hand, simply because that was the way his tribe greeted the bravest of the brave.

In the end the Ambidextral Culture Society turned out to be merely a fashionable hobby. Having led nowhere, except presumably to a certain amount of childhood misery, it faded into the background when more serious matters, such as the First World War, claimed public attention. The society seems to have existed at least until the early 1980s, when in all probability it died a silent death.

12

The Polymorphism of One-sidedness

Although the silent demise of the British Ambidextral Culture Society demonstrates that two-handedness is not necessarily the route to Nirvana, and that it seems impossible to teach by any reasonable means, in one respect the movement's followers did on the face of it have a point. Isn't it the case that, by tradition at least, several of the greatest artists in history used both hands? Holbein, Leonardo and Michelangelo are supposed to have been ambidextrous, for instance, and the famous British painter Sir Edwin Landseer, a close friend of Queen Victoria (who was said to be 'two-handed' herself). The story goes that once, when the subject of ambidexterity arose at a party, Landseer called for two canvasses and two pencils and drew, simultaneously, the head of a deer with one hand and the head of a horse with the other, to the amazement of all present. Whether or not the story is true, clearly some people are capable of achieving roughly the same impressive feats with either hand. Nonetheless, the adherents of the two-handedness doctrine had committed an elementary error of reasoning. The fact that there are people who can perform great tasks with both hands is not at all the same as saying that ambidexterity in itself leads to such astonishing achievements, or to achievements of any kind. There are plenty of people with two hands that work equally well who are not capable of doing anything particularly special with either of them. They don't become famous; instead they have to put up with being labelled 'all fingers and thumbs'.

The most important consequence of the existence of something resembling ambidexterity is that we cannot simply divide human beings into two categories, with a group of pure right-handers on one side and a far smaller group of pure left-handers on the other. A wide range of variation lies between the two extremes. Many left-handed people are aware that there are certain tasks they perform as if they were right-handed. Sometimes they have no choice, for example when using implements designed for right-handed people such as tin openers and corkscrews. In other instances their habit has arisen under duress, as in the case of left-handers who were taught to write with their right hands, or under pressure – sometimes mild, sometimes harsh – from those who brought them up, most commonly perhaps when it comes to table manners. Sometimes things simply turned out that way. Less well known is the fact that the reverse is much more common. The most bizarre case I've come across concerns a purely right-handed man who does just one thing consistently with his left hand: eat. The strange thing here is that table manners are one of the few patterns of behaviour that parents can do a great deal to influence. Holding a spoon with the right hand is the norm, and in many families it's insisted upon from an early age.

Those who don't do everything with either their right hand or their left fall into a category of mixed-handed individuals. It's a group that exists only in statistics, since people never refer to themselves that way in daily life. Most would say they are either right-handed or left-handed and often they are not even aware of any inconsistency in their hand preference, but if we draw a graph showing the degree to which each of a large group of individuals favours one hand or the other, it takes the form of a wave with two peaks: a low one on the left and a far higher one on the right. Most left-handers do most things with their left hand but not everything, and the same applies in reverse to right-handers. Within each of the two groups, only a minority perform all tasks with their preferred hand. Still, there are even fewer who use their two hands more or less equally. People who could be described as entirely mixed-handed seem to be exceedingly rare.

A mixed-handed person is not at all the same, incidentally, as an ambidextrous person, since he or she will use the left hand by preference or exclusively for certain tasks and the right for others. That is a quite different matter from an ability to perform any given task equally well or easily with either hand. The truly ambidextrous can do exactly that, but the question is whether such people actually exist. There are certainly those who claim to be ambidextrous, but it's proven impossible to find people for whom it makes absolutely no difference which hand they use to write, draw, slice bread, peel potatoes, ladle soup or perform other typically one-handed tasks. If they do exist, you might almost suspect that for reasons of practicality they would unconsciously become converts to right- or left-handedness and neglect the other hand. The delights of ambidextrousness must soon pall if every time you want to note something down or put a spoon to your mouth you first have to decide which hand to use.

Of course there are plenty of people who can do certain things using either hand, with equal success and equal ease. This is in fact quite common with relatively simple tasks such as tightening a screw or white-washing a wall. If you work so hard that your preferred hand becomes tired, you can generally switch to the other with a reasonable degree of success. The same goes for working in awkward places, but the hand used in such cases comes into play only when there's a good reason. In normal circumstances, the favoured hand is used instinctively. When it comes to more difficult tasks, such as painting window frames or, harder still, writing and drawing, it's almost impossible to find anyone who can easily switch from one hand to the other.

Michelangelo could, apparently, which gained him a reputation for being ambidextrous. The painting of the ceiling of the Sistine Chapel in Rome was not merely an artistic tour de force, it was a physical one as well. For much of the time he had to lie just under the roof, on high scaffolding. Anyone who has installed a light or mended a hole in the

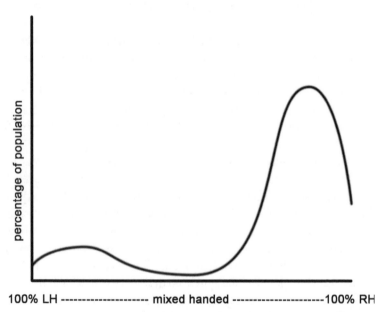

The overwhelming majority of people are either left-handed or right-handed to a significant degree, even if they do not use their preferred hand for every single task. Truly ambidextrous people are extremely rare.

ceiling will have some inkling of what it means to work for long hours, day after day, with your arms raised. Gravity and lactic acid soon make the muscles weak and painful. It's possible the world-famous painting might never have been completed if the artist had not been able to switch from one utterly exhausted arm to the other, but this does not tell us whether Michelangelo was truly ambidextrous. To find that out we would need to know whether he always started the day using the same hand and which of the two he used under normal circumstances, back in his studio. We might not be much the wiser even then, since according to contemporaries Michelangelo was a left-hander who'd been taught to paint and draw with his right hand.

Opinions are deeply divided as to the true proportions of left-handed, mixed-handed and right-handed people. This is mainly because there are almost as many ways of measuring hand preference as there are studies of the phenomenon – and fault can be found with almost all of them. The simplest method would seem to be simply to ask people which hand they favour for certain tasks, but that will render up reliable data only if we can resolve three problems faced by the questioner. Firstly, nice people – which generally includes those willing to take part in research without being offered any significant reward – like to give socially desirable answers. In other words, they try to answer in a way that fits in with their ideas about how they ought to behave, rather than according to the facts of the matter. This tends to produce conformist answers that may not be true.

The second problem appears to contradict the first. Nice people have an equally strong tendency to try to avoid disappointing the poor researcher. They're eager to tell the questioner something interesting, and this tends to produce non-conformist rather than accurate answers. The third problem is that people sometimes have no clear idea which hand they favour for certain tasks and so, entirely in good faith, they answer incorrectly.

It seems as if everything conspires to make the number of true answers as small as possible. This has nothing to do with the subject itself; these are problems that plague all pollsters and they help to explain why political polls can be so misleading. People want both to fit in and to tell the questioner what they think he'd like to hear, and they are only partially aware of their own actions and motivations. As a result, completely unintentionally, they come up with patent untruths. If you think this cannot possibly apply to you, just consider whether at this moment you could say with certainty in which position you are usually lying when you wake up in the mornings, whether you generally cut the toenails of your right or left foot first and which ankle goes on top if you sit cross-legged.

A radical way of avoiding the pollsters' problem is to ask nothing at all but to take objective measurements. One approach has been to measure the strength of the grip in each hand, with the idea that the stronger hand would be the one a person truly preferred. A man called Jules van Biervliet tried an even more subtle approach in 1897. He hung equal weights on the index fingers of both hands of his subjects and asked which was heavier. If an individual said the weight on his right finger was heavier, then he was left-handed and vice versa. Others have used all kinds of methods of measuring the size of the hands and arms, labelling the larger as the one the subject favoured.

This approach allows us to say something about people who are long dead, based on their skeletons. In 1995 researchers at the University of Southampton and English Heritage's Ancient Monuments Laboratory believed they had found evidence that the proportion of left-handed people was slightly higher in the Middle Ages than it is now. They measured the lengths of the bones in the arms of 80 peasants who were buried between the eleventh and sixteenth centuries in the graveyard at the village of Wharram Percy in Yorkshire. In 16 per cent of skeletons the left arm was longer than the right, in 3 per cent both were of equal length and in the remainder the right arm was longer. Based on the assumption that the dominant hand is more often used for carrying loads, and that the right arm would therefore become slightly longer, they concluded there were more left-handed people around in the past than now. The reason for this, they believed, was that in an illiterate world there would be less cultural pressure to be right-handed. It was a story no less muddled than it was entertaining, since cultural pressure has never been about carrying loads but instead relates to things like writing and table manners.

A huge number of carelessly designed studies of this kind have been carried out over the years, all involving the assumption that the preferred hand is the one with which we perform best in certain circumstances: exert the most force, perform the quickest moves, carry the heaviest weights or detect the most subtle distinctions. But whatever they did, the researchers came up with different figures every time, and no wonder, since power and speed are not what hand preference is all about. A few top tennis players aside, our favoured hand or arm is not noticeably stronger than the other, and the cause of hand preference does not lie in the hand or arm anyhow but in the brain, an organ that until recently divulged virtually none of its secrets.

It is therefore more helpful to record behaviour than to look at physical characteristics. After all, the way we behave is the externally visible result of how our brains work. Even this approach is not without its problems, since social and cultural pressure mean that to some extent naturally occurring left-handedness is repressed, as we see even in the tolerant Western world of today when a child learns to write. In a busily scribbling school class there are even now slightly fewer left-handers on show than are present in reality.

This creates a need to look at criteria that are independent of culture, at tasks that are performed with the preferred hand but not governed by etiquette. Back in the 1930s, American Ira S. Wile came up with a truly original solution. He calculated the percentage of left-handers based, among other things, on people he counted as they passed a busy street corner carrying an umbrella or a shopping bag. If they were using their left hands, he categorized them as left-handed. Otherwise he ticked the box for right-handedness. The idea is not entirely absurd, but in practice it was undermined by Wile's inability to control the circumstances under which he took his measurements. A person may carry an umbrella in his left hand because he's left-handed, or it may equally well be because his right arm is tired or aching, because he's injured his right hand, or for a thousand other reasons that Wile could not possibly know about. No wonder he arrived at one of the highest scores of all time: almost a third of the people in Wile's research were left-handed according to his criteria.

If the results are to be valid it's essential to exclude the possibility that people will display a particular behaviour for reasons other than those assumed by the researcher. This explains why psychological experiments usually take place in bare, utterly boring rooms and subjects are asked to carry out minor, often apparently pointless tasks. The fewer external influences at play the better.

Since observing behaviour in controlled circumstances is a time-consuming, complicated and expensive business, most researchers have fallen back on questionnaires. Any number of lists of everyday activities have been compiled, tasks that intuitively seem connected to hand preference: drawing, opening bottles, carpentry and so forth. But each study has used a different list, groups of subjects were by no means always a representative cross-section of the population, and each researcher evaluated and processed the data in his or her own way, so the results remain unreliable and comparison between them is problematic. Furthermore, in most cases little or no account was taken of possible taboos against left-handedness in the subject's home environment. These problems still dog research today, even though standardized lists of questions have been produced, such as the Dutch Handedness Questionnaire compiled by cognitive psychologist Jan van Strien, all of them carefully assessed and judged capable of producing reliable data.

Putting the shortcomings of this motley array of studies into hand preference aside, several firm conclusions can be drawn. Left-handedness is slightly more common in men than in women and twins are rather more likely than other people to be left-handed. Race is not a factor, it seems, even though some studies, mainly from Africa, suggest that in black communities the proportion of left-handers is extremely low. Those particular studies all concern cultures that have a powerful taboo against left-handedness, and the low percentages are not replicated in research among the African-American population of the United States. This raises a further question: how do we explain the fact that a human characteristic occurs down through the centuries, all over the world, in a consistent minority of people of around 10 per cent and slightly more often in men and in twins? Then there is the remarkable fact that time and again we find more left-handers than normal within groups of people with problems and ailments of various kinds, whereas the opposite is never the case. Take a random sample of the population and there's not a single characteristic, circumstance or disorder that can be shown to coincide with left-handedness. Most puzzling of all is the influence of age: the older the people questioned, the less likely they are to say they're left-handed. Before returning to these mysteries we need to look at a question we all too often neglect to ask: what exactly does hand preference involve?

13

The Heart of the Matter

One day in the final weeks of the twentieth century a piano recital took place at the Max Planck Institute in Nijmegen, the Netherlands, and almost all the researchers working there came to listen. Such a frivolous use of precious time was almost unprecedented at the institute, but this was no ordinary recital. Pianist Chris Seed had travelled from England along with his instrument. Seed is a left-hander, like many other pianists, but unlike them he had acted upon a desire to play the piano left-handed. Before he began his performance he explained that this had ultimately resulted in the commissioning of a unique instrument, a pianoforte in reverse, modelled on an early form of grand piano from 1826. It was a faithful copy, except that the long bass strings were on the right while the high notes were hit with the keys on the left.

At first Seed found it far harder to learn to play the instrument than he'd expected. It seemed as if he'd have to begin learning again from scratch. But once he got going, Seed's brain turned out to be perfectly capable of converting everything he'd ever learned into a left-handed playing technique. Exactly what he'd hoped happened: all the pieces of the puzzle fell into place more or less automatically. Seed became at least as good a pianist as he was on a conventional piano and eventually he felt real delight in playing 'as God intended'. The audience, inquisitive and expectant, listened politely and watched with fascination as Seed took his seat at the keyboard.

Then something happened that was in no way exceptional.

Seed played the piano. He played excellently, perhaps even better than on a normal keyboard, although his audience was not in a position to judge. To the initiated it looked a little odd at first, with the pianist appearing to move in the wrong direction, but everyone grew used to that within a few minutes. In every other respect it was purely and simply a piano recital. Why would anyone expect anything different? This was a perfectly normal result, achieved while allowing for a hand preference that differs from the norm.

What then is the preference to which the term 'hand preference' refers? What is the heart of the matter? Of course we feel better and more comfortable when using our favoured hand, and there are many things that with the best will in the world we cannot do the other way around simply on command. But why is that exactly? It's clearly not a matter of strength. Nor is it a matter of ingrained habit, since research using limb movement meters shows that people do not use their favoured hand appreciably more frequently than the other. It doesn't even have much to do with dexterity. A classic test of hand preference, for example, measures how quickly a person can stick pegs into a perforated board. This turns out not to be a good way of telling whether a subject is right-handed or left-handed. Or take a normal right-handed violinist. The nimble dance across the fingerboard is always performed with the left hand, while the favoured hand only has to sweep the bow back and forth. The same applies to guitarists. The left hand dances across the six strings, faultlessly creating the most complex chords, while the favoured hand simply strikes them. Strumming and plucking may be intricate when we play classical guitar, but in pop music they amount to little more than raking the strings with a plectrum. Only in the case of the piano do we see what we might expect: the right hand performs the dextrous melodious work, the left the accompanying, usually simpler bass line.

Nevertheless, a left-handed guitar player will quite often play a guitar that's been stringed the other way, or even an instrument specially developed for left-handers, and most of those who make use of the

Rudolf Kolisch (right) with his ensemble.

alternative options are pop guitarists, like Jimi Hendrix, Paul McCartney and Bob Geldof. With violinists, who generally play in groups, bows poking around in different directions would be downright dangerous. Admittedly, left-hander Charlie Chaplin used a left-handed violin, but he didn't play in an ensemble and he wasn't a professional violinist. The only known case of a professional musician playing a violin with the strings reversed is that of Austrian Rudolf Kolisch, who died in 1978. It does seem, incidentally, that would-be violinists who are left-handed pay a price. No one ever mentions it, but music teachers, when asked, say they 'have trouble with left-handed pupils' and are under the impression that they're more likely to give up. Left-handed professional violinists complain slightly more often about problems with the left wrist, which can become strained.

Among pianists the size and weight of the instrument stands in the way of left-handed playing. They do not usually haul their own pianos around with them. Nevertheless, Seed is not the first to have played a left-handed piano. Back in 1879 the Paris company Les frères Mangeot launched its *piano à claviers renversés*. It was a monstrous lump of an instrument: two grand pianos, one on top of the other, the upper a mirror image of the lower, put together in such a way that the two keyboards were stepped back like those of a church organ. It could therefore serve as a left-handed piano, although the main idea was that the pianist would no longer have to make great leaps from one end of the keyboard to the other and could use the same fingering for either hand. It was not a success. Since 2001 the renowned German firm Blüthner has been manufacturing its Model No. 4, a left-handed upright piano, but such instruments are still extremely rare.

The *piano à claviers renversés*, built by Les frères Mangeot.

Even though the vast majority of left-handed pianists play in exactly the same way as their right-handed colleagues, here lies the key to what hand preference really means. There's an enormous difference between what a pianist is called upon to do and what a violinist or guitarist does. A pianist uses both hands to do roughly the same thing: strike keys. It's usually necessary to strike them harder and more quickly at the high end of the sound spectrum, which is why a normal piano is designed so that the right hand has the more difficult job, but the point is that in theory the two hands work independently of each other. Were that not the case, Maurice Ravel's *Concerto for the Left Hand* could never have existed. Ravel was commissioned to write the piece by a brother of the world-famous philosopher Ludwig Wittgenstein, Paul, after he lost his right arm in the First World War. It has to be said: with the help of Ravel's genius it's astonishing what a one-handed pianist can achieve.

Had Paul Wittgenstein been a violinist or guitarist, no concerto would have been written for his left hand. He would have been unable to produce a simple children's song, even a single note. Such instruments require the two hands to perform quite different tasks, complementing each other like Yin and Yang, combining to create the sound.

The positions of the fingers on the fingerboard of a stringed instrument are not greatly different from the fingerings on the keyboard of a piano, which every pianist learns to perform well with either hand, after a little practice. The fact that in the case of the violin and guitar we always leave the job of fingering to the non-favoured hand shows that in our apparently simple strumming and bowing lies something that makes considerable demands. It must have to do with rhythm, and with having complete mastery of the dynamic. It's not a matter of strength but of control, of minimal, extremely precise differences in the force, pressure and speed of movement. Strumming and bowing involve exceedingly subtle steering.

With this in mind it becomes clear why people perform other tasks in which control and precision are important better and more easily with their favoured hand. Snooker players, for example, use their preferred hand to strike the ball, even if they have to tie themselves in knots to do so. The reason is not that they are required to hit it hard but that they must shoot with supreme accuracy, not only aiming perfectly but using precisely the right degree of force and ensuring that the cue doesn't move so much as a millimetre too far. The ball must be knocked forward, not pushed away. Even when doing something as banal as sweeping the floor, we almost all use our non-favoured hand, low down on the broom-handle, to achieve a relatively crude thrusting motion, while the favoured hand is held loosely at the top to guide the broom and pull it back at the right moment.

Hand preference, then, is a subtle matter, even though we experience it as quite compelling. That compulsion is not absolute, since few people perform every single job that requires subtle guidance with their favoured hand, and when called upon to do so they can generally manage almost any imaginable task with the other. Musicians are a good example. If anything, orchestras include a disproportionately large number of left-handers, but they almost all play their instruments in the standard manner, as if they were right-handed. Left-handed people who are forced to learn to write with their right hands may encounter all kinds of difficulties, but they generally learn to write just as well as anyone else, and people whose preferred arm is put out of action temporarily or permanently, for whatever reason, quickly grow accustomed to doing everything with their remaining 'good' arm, including complicated tasks like writing. They too generally manage so well that after a while it's impossible to tell that there's anything special about them.

If necessary – and no necessity exists in the case of a healthy left-hander learning to write – you can escape your natural hand preference for almost all purposes and in virtually any situation, given enough training, sometimes even to the extent that you can reach the top internationally, as left-handed cellist Pablo Casals did by playing the way any right-hander would. Only where a slight diminution of reliability and control might have serious consequences is it better to forget about switching hands. It's no doubt advisable to let the dentist and the surgeon use whichever hand they prefer.

14

The Power of Small Differences

In the main, living conditions in the army used to be abominable: treatment was appalling and remuneration paltry. Given half a chance, almost everyone would try to avoid military service. In the period when lots were drawn this was not particularly hard, as long as you had enough money. You simply hired an impoverished replacement. The barracks were full of men too poor or too stupid to have found a way out, in many cases uneducated peasants who had never been more than a few kilometres from home. Conscripts could barely be taught even the simplest exercises, since most were unable to tell right from left and therefore had no idea how to put their best foot forward. There was one effective solution to the problem: a tuft of hay was placed in a recruit's left boot and in the right a tuft of straw, materials with which all young country lads were familiar. The sergeant no longer shouted 'left – right' but instead 'hay – straw'. In no time this approach transformed a motley bunch of blundering young men into a shipshape platoon on the march.

It's such a good story that, to quote Ethel Portnoy, it must be a monkey sandwich, in other words a modern legend, a story everyone has heard but no one has witnessed and for which no one can provide a precise date or location. Nevertheless, the anecdote has spread far and wide. The Dutch version usually makes them soldiers in Belgium or the rural province of Limburg, the English tend to enjoy telling theirs as a tale about recruits from the Scottish Highlands, in America it features the tsarist armies of Russia, and in France the bumpkin soldiers are Corsicans or units commanded by the German Emperor. As with many such legends, there's at least a grain of truth in it: people do tend to get their left and right mixed up, whereas they don't have the slightest trouble telling other opposites apart, such as top and bottom.

The tricky left–right distinction is nevertheless crucial in many fields of life where we might least expect to encounter it. Of course everyone knows that left and right are important in traffic and in reading and writing, but there are many other ways in which they have an important if subtle part to play in our perception, interpretation and experience of the world around us.

Some brain researchers claim that thoughts and abstract concepts are a kind of mental simulacrum of bodily experiences. This hypothesis, known as embodied cognition, is less peculiar than it may seem. Behind it lies the idea that the only stimuli to reach the brain from outside are signals from the body, and some of those signals have their origins within the body. This certainly holds true for all kinds of information about the internal condition of body parts. Other stimuli do ultimately have external origins, but no direct contact is possible between the brain and the outside world. Instead, external events first stimulate one of our senses, which in turn sends a signal to the brain. The brain can experience only corporeal events, so what does it ultimately use as the building blocks for thoughts and concepts that have nothing to do with the body? Exactly. Those same bodily experiences. There's nothing else available. In which case, the adherents of this line of reasoning argue, people with bodies that are constructed in a different way must think differently about certain things.

In 2009 Daniel Casasanto, of the same Max Planck Institute where Chris Seed had given his piano recital a decade earlier, found indications that this is indeed the case. In a series of tests it turned out that right-handers associate the space to the right of them with positive values such as 'good', 'pleasant' and 'successful' more readily than the space to their left. With left-handers precisely the opposite applies. It's as if the preference for one hand over the other radiates out into the vicinity of that hand. This means for example that the same portrait photo, when placed on a table to the right of a right-hander, will be seen in a more positive light than when it happens to be placed on the other side. It may even mean that when an employer looks at a list of brief descriptions of job applicants that has been laid out in two columns, those in the column on the same side as his or her preferred hand will be judged more favourably. If this turns out to be true, then perhaps elections, selection procedures and recruitment are even less rational processes than we already feared. It seems there isn't an awful lot we can do about that.

There are other ways too in which the relationship between left and right determines our take on reality. We like symmetry, a balance between left and right. In the environment of our daily lives we see this clearly enough. Chairs, tables and most of the other objects we use tend to be symmetrical. We place an equal number of chairs on either side of a dining table, directly opposite each other, since the whole set-up looks messy otherwise. This doesn't mean we can't appreciate asymmetry in furniture and utensils. We're delighted if an accent breaks up the symmetry and by doing so emphasizes it – for example if we place the tablecloth at a carefully calculated slant and the vase of flowers deliberately off-centre. But asymmetry generally shouldn't be taken any further than this. Asymmetrical furniture is strictly for arty-farty people eager to demonstrate how eccentric and non-conformist they are. An asymmetrical espresso machine is an example of radical design, something for the connoisseur who wants to show he's averse to bourgeois consumerism and therefore chooses products with an industrial look, machines whose beauty does not rely on the crude satisfaction of our desire for symmetry but instead on functionality. Beauty that arises out of a natural disregard for considerations of beauty on the part of engineers, who are fixated on utility – that's the kind of asymmetry we find stimulating.

It's different of course with images that, rather than presenting functional items, depict events, situations or scenes from nature. A landscape or setting is rarely symmetrical, yet even here left–right symmetry has an effect. In contrast to the pairs top–bottom and back–front, symmetry between left and right produces a sense of balance and calm. One obvious example is the deliberate, overt symmetry of the classical, artificially clipped and trimmed French palace garden. In a more subtle, more concealed way, the English version of large-scale formal planting, the landscape garden, derives much of its attraction from symmetry. It's this that gives the artificial landscape its aura of cosiness and safety, while carefully positioned asymmetrical elements help it to avoid the static dullness that attaches to the French style. Asymmetries create tension and unease, accompanied by a sense of dynamism. The panorama of a formal garden in some sense creates a story, which you read by looking around. A French palace garden is like a tiled wall, featuring endless repetition.

Differences between what we see to the left and to the right affect, almost imperceptibly but to a significant degree, what we see in a painting or diagram, on television, or on a cinema screen or computer monitor. Artists and designers take due account of this, even without being aware of it. They have to, because the distinction between left and right has an impact on our perception at every imaginable level. It's an unavoidable aspect of how the natural world appears to us, and of the physical architecture of our visual faculties. The influence of cultural aspects such as the direction in which we write makes itself felt too, as does the fact that the vast majority of painters are right-handed. Nevertheless, we should not forget that our ability knowingly to distinguish between left and right is in itself quite remarkable.

15

How Freud Found his Right Side and Pooh Didn't

In *The House At Pooh Corner*, A. A. Milne describes Winnie the Pooh failing to learn to tell his left from his right: 'Pooh looked at his two paws. He knew that one of them was the right, and he knew that when you had decided which one of them was the right, then the other one was the left, but he never could remember how to begin.'

It's the same as the problem we encountered in the illiterate recruits with their hay and straw, and something that children find harder than adults. Sigmund Freud, founder of psychoanalysis, vividly remembered how he had thought of a trick when he was a child slightly older than Christopher Robin: he inconspicuously pretended to write something down. The hand that automatically started to move must be his right, so then he knew, for a while at least, which side was his left. Innumerable children have thought up solutions along these lines, sometimes guided by a birthmark on one hand, a bracelet or some other clue.

Although adults find it a good deal easier to avoid confusing their left with their right, they're by no means perfect at it. A great many motorists have unthinkingly turned right when asked to turn left. Computer programmers know only too well how easy it is accidentally to confuse the symbols '<' (smaller than) and '>' (larger than) in lines of programming and how difficult it can be to track down this simple error. Unpractised writers may reverse the occasional letter; in fact, this happens quite regularly when texts are written in giant script, as on notice boards listing the day's special offers at a butcher's or greengrocer's shop. It seems that when a sign-dauber has those big letters right under his nose he can lose track of what he's doing.

As long as we have a clear, obvious reference point, such as Freud's writing hand, we're capable of overcoming any confusion, but in situations lacking such indicators we have more difficulty. Which, for example, is the left side of a stage? It depends whether you're looking from the stalls or from on stage. We need to make arrangements that will ensure instructions like 'enters stage left' are open to only one interpretation. The French have a neat mnemonic. Just as on board ship right is starboard and left is port, the French stage has its own names for right and left. Seen from the auditorium, the left side is called the *jardin*, the garden side, and the right the *cour*, the courtyard side. Spectators can remember which is which by thinking of the initials for *Jésus Christ*: the J for *jardin* is on the left and the C for *cour* on the right. Meanwhile an actor can remember the difference because of the similarity in sound between *cour* and *coeur*, or heart, which beats just to the left of centre. So the actor can move to the left backstage and emerge, as far as the audience is concerned, from the right without anyone getting into a muddle.

Our uncertainty in these matters is far more likely to play tricks on us when we have to choose between left and right than when we're faced with opposites such as front–back or above–below. Ask a hundred people to take a writing pad out of the left side of a cupboard and there will always be a few who peer along the shelves to the right, but no one will kneel down if he's asked to fetch something from the top of that same cupboard. When children learn to write, they can easily mix up the d and the b, whether they're right- or left-handed, but only in exceptional cases do they mistake a p for a b. The difference between top and bottom, just like the difference between front and back, is far more obvious than the difference between left and right.

It's exactly the same with animals, except that most have infinitely more difficulty with the left–right distinction than we do. They can easily see that a threat comes from the left or the right, or that there's something good to be had on the left or the right, and respond accordingly, but what they rarely if ever manage to do is to link a choice between left and right to a stimulus that in itself has nothing to do with that distinction.

Suppose for example that we put a rat, a pigeon or some other creature into a cage with a button in it marked with an arrow that looks like this: '>'. If the arrow points to the left, then pressing the button produces a tasty morsel of food, but if the arrow points to the right then something unpleasant happens, such as an electric shock. The bird or animal's task, therefore, is to learn to see which way the arrow is pointing and thereby deduce whether or not it's advisable to press the button. Hardly any animals have proven capable of mastering this skill, but the results are quite different if the arrow points either upwards or downwards. Rats, pigeons, apes, even octopuses can learn to respond reasonably effectively in that situation. Up or down seems to be a clear and recognizable distinction, whereas in the eyes of an animal '<' is indistinguishable from '>'. For a while it was thought that pigeons were an exception, until it turned out that those particularly cunning birds were tilting their heads as they looked, transforming the left–right distinction into an up–down one.

Our blind spot for the left–right distinction has to do with the fact that the difference between left and right has virtually no role to play in the natural world, or at least those parts of it that are visible to the naked eye. Above and below differ as strikingly in everyday reality as they do in essence. Birds of prey menace you exclusively from above, never from below. You dig holes in the ground, never in the sky. The distinction between front and back is no less fundamental: if you have to flee, or indeed if you want to catch something, then you need a clear idea of which is your front, otherwise you won't get very far. This does not apply to left and right; in fact it may actually be a disadvantage to see the two sides as fundamentally different. After all, the same enemies might lurk on either side and the same desirable things might be found to our left or to our right. Anyone who, having already been attacked from the left, recognizes a danger to the right as 'the same' – and therefore quickly decides to escape by moving in the opposite direction – has a clear advantage over someone who sees a threat from the right as an entirely new phenomenon that must be judged on its own merits.

The left–right distinction is insignificant when it comes to the structure of plant and animal species, whereas the front–back distinction is usually quite clear, and the difference between top and bottom almost always. In the left–right plane, almost all organisms are symmetrical, with the exception of some species of shellfish. Their shells have an asymmetrical form, in most cases spiralling one way around but sometimes the reverse, although that makes no difference to the species. It's a meaningless variation. We see a similar thing in flatfish: sometimes one side, sometimes the other grows into the lower half.

It's a very different story in the world of the invisibly small, at the level of atoms and molecules. Molecules that are each other's mirror image have completely different characteristics for that reason alone. Outside the microscopic world, which by definition we cannot observe without special devices, the left–right distinction is an important feature only of things made by man. An ability to distinguish between an image and its mirror image is required exclusively for items of cultural significance. This becomes clear if you compare the mirror image of a landscape photograph taken at random with the mirror image of a photograph of a typical man-made environment, such as a shopping street. With the first photo it's far from easy to make out whether you are dealing with the original or the mirror image, but with the second it's immediately obvious.

The alphabet is one of the cultural artefacts in which mirroring is a prominent factor. S, Z, R and N are simply incorrect if reversed (unless explicitly assigned significance in some way), but several letters of the alphabet are transformed into others if we look at them in the mirror. For example, p and q are left-right mirror images of each other, as are b and d, while p and q have the same shapes as b and d upside-down. If we reverse an n both horizontally and vertically we get a u. The uses of mirror-sensitivity are not limited to asymmetrical shapes like letters. The traffic rules, in which the distinction between left and right is a mainstay, is a wonderful example of perfect symmetrical mirroring. Whether you're driving from Aberdeen to Yeovil or from Yeovil to Aberdeen, your image of the traffic is exactly the same.

Despite the subtle games we can play with mirror images, our memory is still set up in a way that coincides with the natural state of affairs – the animal within us. When we store away the images we see, we give low priority to the retention of information about their left–right orientation. This is clear from experiments that resemble a well-known parlour game. People are given a large number of random pictures to look at. A little later they are again presented with a series of pictures, some the same, some different, some in mirror image and some upside down. Time and again people prove able to recognize pictures they have seen before but in mirror image. They don't even notice that the picture has been reversed. They are far less likely to recognize images that have been turned on their heads, especially in the case of abstract shapes; at least, they fail to give the appropriate response when asked. Mirror images of depictions that are stored in our memories do not stand out as different, whereas those reversed vertically do. No wonder children learning to write have such difficulty remembering the difference between d and b, while the distinction between p and b rarely causes them any trouble.

The way our bodies are constructed fits neatly within this general rule. The difference between our fronts and backs, or between our top and bottom halves, is significant and profound, just as in all other vertebrates, but the difference between left and right can barely be seen at all externally. On both sides we have an eye, a hand, a foot and a great deal else, and body parts of which we have only one, such as noses, navels, penises and vaginas, with all that goes with them, are right in the middle and themselves more or less symmetrical. The only exception is the parting in our hair. Plus natural blemishes such as birthmarks and warts, but then that's what makes them blemishes.

Nevertheless, it's because of the incompleteness of this apparent symmetry that we can tell left and right apart at all. If we were perfectly symmetrical, we would have no way of distinguishing between that which lies to our left and that which lies to our right. Our own mirror image, for example, would be exactly the same as our real-world image, so we would be unable to tell the difference between our actual appearance and the way we look in the mirror. As a result we would simply not notice that our image had been reversed. Nor would we notice if the entire world around us suddenly switched, as happened to Lewis Carroll's Alice when she stepped through the looking-glass. Any experience we perceived on our left side would in no way differ from the same experience to our right. So we'd be able to see that a d and a b were each other's mirror image if they were written next to each other on a piece of paper, but we'd never be able to explain how a d should be written. Freud's childhood memory of having to work out which was his writing hand illustrates the point perfectly. When he no longer knew which was which, he made himself extra-asymmetrical by putting his writing hand to work.

The fact that children have so much more difficulty telling right from left could have to do with the fact that adults are less symmetrical than children. Like the rest of our bodies, our brains grow during childhood, not only in size but internally. Their structure changes.

The brain of a newborn baby is to some extent comparable to a recently built office block. The basic facilities are in place but as yet it's unfurnished. On completion of the structure, all the rooms are interchangeable concrete spaces, but within a few months every part of the building has been occupied by one department or another. The third floor, for example, may have become the financial heart of the company, while the canteen is on the first floor; the left side of the fifth floor houses public relations and to the right of the lift are the sales staff. In the process, cupboards have been moved several times, desks turned around, extra lamps brought in. Things that didn't appear in the initial plan for the layout have turned out to work better in practice. The way a building ends up is therefore determined partly by the initial blueprint and partly by a learning process, by trial and error.

In the same way, parts of our brains are furnished according to a standard plan that's anchored in the genes of every new world citizen, while others develop according to a learning process generated by the external influences operating upon the child. Over the years a vast number of new connections are made between neurons and a good many existing connections disappear, so that ultimately a set of circuits emerges that can get us through adult life successfully.

In this sense the development of the human brain does not differ greatly from that of other mammals. Their brains too are incomplete at birth; they too need to experience the world before they fully reach adulthood. In humans the process is longer and more complex, but there's something else as well, something quite special. Human brains differ from those of other mammals in the size of the upper, outer layer, the cerebral cortex. It's there that the so-called higher functions are located, including things we regard as typically human. In other mammals the cerebral cortex is symmetrical, consisting of two roughly identical halves connected by a broad bundle of axons called the *corpus callosum*. Only in humans, or at the very least in humans far more than in any other animal, some parts of the cortex that specialize in specific tasks are found on one side only. As a result the two halves of the brain, though largely identical at birth, eventually come to feature significant differences. They may look symmetrical, but adult human brains work asymmetrically to some extent. Many functions that have to do with speech are generally found in the left half, as are arithmetical skills, while the right half tends to be engaged with the final processing of visual and spatial sensations. The right side of the brain commands the databank of faces that ensures we don't simply walk past family, friends and colleagues on the street without noticing them. People also somehow become able to control one hand better than the other. Because of this typically human process of one-sided specialization, known as lateralization, the brains of adults are far less symmetrical than those of small children.

16

Why a Running Rabbit Doesn't Tear Itself Apart

Although it's harder for us to tell left from right than, for example, top from bottom, we can recognize mirror-image likenesses between left and right much more easily than other symmetries. That's only logical. Your life may depend on it, whereas in emergencies chance symmetries between top and bottom or front and back are merely a distraction. What happens off to our left requires a reaction that's a neat reversal of the way we react to precisely the same event to our right, whereas what happens in front of us demands a totally different reaction from what happens behind us.

Our sensitivity to left–right symmetry is clearly visible in the things we make. Traditionally, cathedrals and other large, official buildings are symmetrical when seen from the front, just as we are, but when looked at from the side there's no symmetry between the front and back halves. Such buildings, intended to be seen mainly from the front, have to radiate certainty and authority. The sense of repose evoked in us by symmetry helps them to achieve this. The same does not apply to fairytale castles, which, from Neuschwanstein in Bavaria to the Queen's castle in the Walt Disney film *Snow White and the Seven Dwarfs* (1937), are intended to create romantic tension, an enjoyable sense of mystery and adventure. They are therefore anything but symmetrical.

Friezes and decorative paintings often feature some form of left–right symmetry, but only rarely do we see a mirroring of top and bottom, the way riverbanks are reflected in the water. Clearly we are particularly sensitive to left–right symmetry, but how we detect it is another question, one we can answer only in part.

By and large our visual system consists of the retina deep within each eye, the visual cortex and the optic nerve that connects the two. The visual cortex is at the back of the brain, so although we look with

At first sight the lower of these two patterns by Bela Julesz looks symmetrical, the other asymmetrical. In fact they are the same, except that one is perpendicular to the other. If you turn this book through 90 degrees, the symmetry will move to the other pattern. We immediately notice left–right symmetry but are less likely to spot top–bottom symmetry.

our eyes, we see with the backs of our heads. The eyes appear to be connected crossways to the left and right halves of the visual cortex, in the same way that many other body parts are controlled by the opposite half of the brain, but it only seems that way. In reality the system is more refined: the right half of the brain processes information coming from the left half of the field of vision, which falls on the right half of the retina of each eye, while the left half of the brain deals with information from the left half of each retina, which comes from the right half

Neuschwanstein, the fairytale castle built by the romantic but shy King Ludwig II of Bavaria in the second half of the nineteenth century in 'the genuine style of the old German knightly fortresses', as he wrote to Richard Wagner, 'with the winds of heaven blowing around it'.

of the visual field. The result is that the right half of the brain processes everything that happens on the left side of whatever we're looking at, and vice versa.

This may strike us as an unnecessarily complicated way of going about things, but the evolutionary advantage it confers is easy to recognize. Eyes have to be exposed to the outside world, otherwise they would see nothing, but this makes them more vulnerable than the brain, which is stored safely away in the skull. The layout of our visual system ensures that even if one eye is put out of action completely, we can still use our entire visual cortex, since each eye serves both halves of the brain. A one-eyed person has a larger blind angle than normal and can no longer see depth – this applies only to the handful of species that have two eyes facing forwards – but in all other respects their visual faculties remain just as good as anyone else's. Were both eyes connected in their entirety to one cerebral hemisphere or the other, the loss of one eye would partially or entirely disable half our visual brain capacity.

All mammals have the same mishap-resistant set-up, but not all vertebrates. Pigeons, for example, have a system in which each eye is connected exclusively with one half of the brain.

This raises the question of how our visual system manages to recognize left–right symmetry. We cannot say for certain. A number of suggestions have been made, the most credible being an idea put forward by Bela Julesz, a researcher at Bell Laboratories in America, in about 1970. His theory, which builds on earlier work by the Austrian physicist and philosopher Ernst Mach, goes roughly as follows.

The visual cortex of each cerebral hemisphere includes among other things an area in which information coming from the eye is 'pictured' before being processed further. We will call this area the projection screen, even though that's a misleading term in some ways, since of course no one is actually looking. To put it rather simplistically, the stimulation of the projection screen area by signals from the retina is the first phase of the process we call seeing.

If we look at the centre of a shape that has left–right symmetry, then one half of it is pictured by both eyes in the left side of the visual cortex and the other half in the right side. To recognize symmetry we have to compare every point on the left half of the projection screen with its corresponding point on the right. This is made possible by the *corpus*

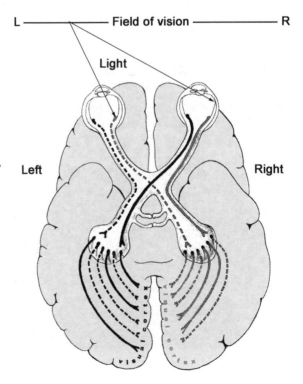

The visual system, in broad outline. Light entering from the left of the visual field travels to the right side of the retina in both eyes, and from there signals travel to the right side of the primary visual cortex at the back of the head.

callosum, a broad bundle of some 150 million axons that connects the two halves of the brain. If almost all the points we compare are stimulated in the same way, then we know we're looking at something symmetrical.

Julesz produced symmetrical patterns that provide quite strong evidence that the detection of symmetry does indeed essentially work this way. If, after identifying symmetry, we then focus not on the middle of the pattern but on some other point to the left or right, we are suddenly unable to see the symmetry any longer. What's left is an unstructured jumble of black and white squares.

It seems a sound argument, but there is one serious snag. How come we're capable of making comparisons of this kind? The answer is by no means obvious.

Seeing is an involuntary process, which cannot be switched off even if we deprive ourselves of sight by shutting our eyes. Anyone who has ever sunbathed on a beach knows that we can't ignore the light that penetrates our eyelids. Everything that gets through to our eyes from whatever is happening in front of us is sent to the visual cortex, which puts together as coherent an image as possible, whether we want it to or not. This is not to say that we are conscious of everything we see. On the contrary, most of the signals that reach us are dismissed as unimportant

The same symmetrical Julesz pattern as before. If we focus on an imaginary line between the two arrows, we suddenly cease to see any symmetry.

after passing through an initial stage in the process. Sometimes they get stored away in the memory, but often they are immediately forgotten.

The degree of relentless diligence the visual cortex exhibits in processing images at lightning speed all day long without our noticing is demonstrated by the reflex reactions we are capable of having to things in our visual field. A fly or a drop of fat from a frying pan heading straight for an eye makes us close the eyelid with remarkable speed. If we notice it at all, it's not until afterwards. We would never be able to react consciously as quickly as that. Everyone knows the feeling of having seen 'something' but not being able to say quite what it was. It's a popular feature of detective novels.

Seeing consists of a great deal more than simply the depicting of impulses on the projection screens of the primary visual cortex. It's important to remember that on each projection screen, images from each eye relate to only one half of the field of vision. First the data from both eyes has to be compared and combined, so that a coherent half-image is created (one that has depth as a result of the slight difference in the angle of vision between the two eyes). Then those two half-images have to be linked together into the one, seamless, complete image that we experience.

Seeing symmetry is as involuntary as the act of seeing itself, so it's reasonable to assume it's an integral part of the process of comparison and splicing together of half-images. In other words it's apparently here that mirroring enters into the process in some way. But this presents a significant problem, since everything suggests that we do not use mirroring in interpreting signals coming from the retina. If we did, we'd become severely disorientated. It would be like living in a crazy fairground attraction.

Imagine what would happen if the splicing together of the two half-images, in other words the integration of images in the two halves of the brain, involved mirroring. A rabbit running past would appear to run back towards where it came from as soon as it passed the centre of our field of vision. This is clearly not the way it works. If it were, not only would we hardly ever catch any rabbits, but we'd also be terribly accident-prone. It's even less likely that the mirroring takes place as we compare versions of the same half-image from our left and right eyes, in other words within one and the same cerebral hemisphere. If it did, then a rabbit that according to our left eye was running from left to right would be running in precisely the opposite direction according to our right eye. This would also make it completely impossible for us to detect symmetry, since after mirroring, none of the points in the image from one eye would coincide with its counterpart in the image from the other eye. Seeing depth, a capacity that relies on slight differences in position of the same impulse on the two retinas, would be impossible as well.

So again we find ourselves confronted with a paradox. To see symmetry we must apply a process of mirroring, but that would make it impossible for us to see anything properly. Fortunately, Julesz came up with a solution. He looked at the vast majority of animal species that do not have both their eyes on the front of their heads but one on each side. Looking sideways, they are forced to create mirror images, and they need to do this at the stage when they're interpreting the half-images from their left and right eyes. It goes like this.

If a butterfly flies past the left side of a rabbit from its back to its front, then the image moves across the retina of the rabbit's left eye 'from nose to ear'. If the butterfly passes to the right of the rabbit, then the right eye experiences precisely the same thing. Nothing unusual in that, it would seem. But when they arrive at the projection screen in the rabbit's brain, the two movements, which were in the same direction, appear to run counter to one another. To allow the rabbit to experience the actual direction of movement in both cases, a correction has to take place somewhere, and it must take the form of mirroring. It becomes clear how important this is if we have the rabbit move as well, running so that the world moves in relation to its eyes. If the animal interpreted the images coming from its two eyes without mirroring, then it would be torn apart inside by the firm conviction that one half of its body was running backwards at the same speed as the other half was moving forwards.

Parallel eyes on the front of the head are the exception, and they may be a fairly recent evolutionary phenomenon. If we go far enough back in time, at least a few tens of millions of years, then we come to our own distant ancestors, who like the rabbit had eyes on the sides of their heads. They must have had a visual system suited to such an arrangement, and here may lie the explanation for our sensitivity to symmetry. We mainly function according to a visual system that is relatively new and does not involve mirroring. Although it is adapted to our parallel, front-facing eyes, Julesz argued that it still contains echoes of characteristics of the much older system that did involve mirroring. It's possible that this system is so old that rather than being located in the cerebral cortex it's lodged in the brain stem, which is far more ancient in evolutionary terms – a legacy of the immeasurably deep past in which we still communed with crocodiles, and a feature that now proves fantastically useful.

Julesz's theory is an admirable attempt to find a solution to the mystery of how we recognize symmetry, but it's probably not the final word on the subject. It has too many weak points for that. For example, the mirroring capacity of animals with eyes on the sides of their heads neutralizes an apparent contradiction between partial images within one half of the visual field, whereas the detection of symmetry concerns the connection between the two halves of the field of vision. Furthermore, the fact that we cease to experience the symmetry of Julesz's patterns if we focus on a point well away from the axis of symmetry proves that mirroring to detect symmetry has nothing to do with the way rabbits see. To rabbits, butterflies travel in the correct direction even if they are at the very edge of the field of vision, quite apart from the fact that sideways-eyed creatures like the rabbit have no more to gain than we do from a process of mirroring the two halves of the visual field.

As if that were not bad enough, there are at least two other important phenomena about which Julesz's theory has nothing to say. One is the fact that although in looking at complex patterns we recognize symmetry only if we concentrate our focus at or near the axis of symmetry, simpler symmetries such as those of a baroque vase or a person's face are always obvious to us. The second point is that we can recognize symmetry between top and bottom, although it takes us slightly more effort.

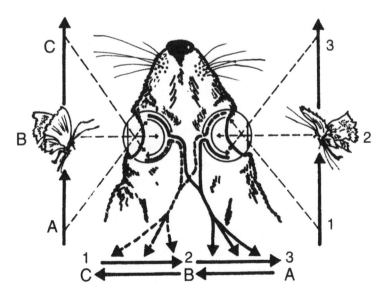

Were it not for mirroring, two butterflies flying in the same direction (from A to C and from 1 to 3) would appear to the rabbit to be moving in opposite directions.

Perhaps the pigeons described earlier can help us along here. The birds gave a false impression of being able to tell left from right, or to be more precise to tell the difference between two adjacent buttons marked '<' and '>'. In reality they had transformed the left–right distinction into a manageable one between top and bottom by tilting their heads to one side. Their approach was to rotate the image their visual system had to process such that it became meaningful. Now of course we must not forget that the pigeons were attempting to differentiate between the two halves of an image, whereas the detection of symmetry requires us to search for similarities. That needn't worry us, however. Symmetry is after all a form of distinction. It has to do with the systematic difference that we all recognize between an image and its mirror image. Anyone who cannot tell left from right, and therefore cannot distinguish between '<' and '>', will also be unable to detect left–right symmetry.

It's conceivable that we do something similar to what the pigeons do, only inside our brains. To put it very simply, it may be that our visual system can rotate images on the projection screen and then compare them. This suspicion is reinforced by the fact that we must be able to shift images around, since that's how we measure the difference between the images from our two eyes that gives us our perception of depth. We know this is how depth-perception works, because based on the same principle we can create an experience of depth in photographs, films and television images by setting up two pictures of an object so that they are slightly offset and presenting one to the left eye and the other to the right. This is usually done with the help of a pair of glasses with differently coloured lenses, each of which filters out one of the two images.

There are further indications of our mental ability to rotate, shift, even bend and deform images. Things need to get pretty extreme before we find ourselves making any real effort to recognize an everyday object such as a chair, irrespective of the position from which we view it. Even if we see a chair from a given angle for the first time, so that we can't precisely match what we're looking at to an image stored away in our memories, we immediately know what we're seeing. That's only possible if our standard, memorized image of a chair (whatever that may look like) and the image on the projection screen can be turned and moved around until they more or less match. When we succeed in doing this, we conclude: 'Ah, a chair.'

In recognizing symmetries, the memory and the mental mechanisms that interpret images and store them in our heads as distilled concepts are probably of great importance. Along with our ability to shift images around, this would explain why we have no difficulty recognizing symmetry in any more or less familiar object, even if it's on the edge of our field of vision. It's a different story when we're dealing with something we've never seen before, never filed in our memories, especially if it's complex and bears no resemblance to anything else. Then our memories are no use to us and we have to resort to mirroring. Julesz managed to produce exactly such uninterpretable, random patterns, and as he was developing them he took care to throw away all test versions that accidentally included something recognizable, the way we see faces in clouds.

Julesz's random-dot patterns are examples of the most difficult kind of symmetry: extremely complex and lacking prior models or clues. As a result, if we focus far from the axis of symmetry the visual system is poorly equipped to compensate by shifting the image around. It's conceivable that in such cases it makes a quick check and finds no reason to look for symmetry. The opposite applies, it would seem, with familiar, simpler, or more easily reducible shapes.

The same mechanism might also explain why we have more difficulty with top–bottom symmetry, even though we're quite good at recognizing it. Apparently we have to do a piece of extra, non-routine work: rotation.

All things considered it seems there's a link between our capacity to tell left from right and the trick of detecting left–right symmetry, and both have to do with our capacity to perceive depth. This raises a suspicion that the experience of symmetry and the ability to distinguish left from right may be confined to those animals that, like us, have both eyes in the front of their faces. Among land animals, this group is quite small, consisting only of species that rely heavily on sight. They are mainly apes, other primates and felines. Those with eyes to the side generally have less good eyesight and rely more on hearing, smell and touch. It's therefore none too surprising that such animals are unable to tell left from right. We may indeed wonder whether they can detect symmetry, and even whether they can recognize a chair automatically from the most peculiar of angles as we can. No shortage of conundrums here.

Birds, finally, depend so much on their eyes that they must surely be able to perceive depth. To dispel any doubts on the subject you have only to watch a buzzard snatch a mouse from long grass with deadly precision. Yet most birds have their eyes very much to the sides. It seems their brains have a mechanism that can judge depth based on the independent images from their two eyes or – and this may be a better guess – from a lightning-fast series of images from the same eye. That would work as long as the bird was flying or running, or at least moving its head.

17
Tintin's Law

Few worlds have been as reliant on superstition and ritual as that of classical theatre. One golden rule in the Western theatrical tradition concerns the entrance of the messenger. A bearer of good news would enter stage left, as seen from the stalls. If he made his entrance on the right, then you could bank on bad tidings. The direction in which the messenger moved therefore indicated to the audience what was about to happen, and spectators were subconsciously aware of that implicit message.

Few messengers put in an appearance on stage these days, but that certainly doesn't mean that golden rules are a thing of the past. In fact there are now more of them than ever, as a result of the vast array of technical facilities that are now available. Modern media such as films and comic strips stick to them as firmly as the dramatists of old did, even if audiences have no intimation of their existence. The same goes for painting, even photography. Much of what we generally call 'composition' and regard as artistic seems in reality to be determined by simple but deeply ingrained rules that dictate how we interpret images. Those rules ensure, among other things, that we detect a direction in everything that is the least bit suited to pointing one way or the other.

Among the most convincing examples of this are the graphs we see every day. We involuntarily assume that their lines begin on the left and end on the right. If the right end is higher than the left, then we perceive the line as rising, whereas if the right end is lower we assume the line is falling. We're so inflexible in this regard that no business-man in his right mind would try to illustrate his company's annual report with graphs that work the other way, even though they would contain precisely the same information.

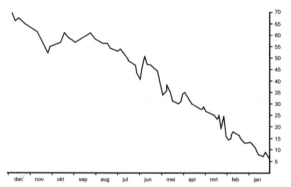

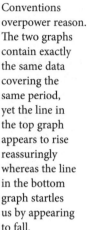

Conventions overpower reason. The two graphs contain exactly the same data covering the same period, yet the line in the top graph appears to rise reassuringly whereas the line in the bottom graph startles us by appearing to fall.

Rising and falling are actually rather strange concepts in this case, since lines on a page are inert. They don't actually move an inch. Nonetheless we perceive movement. We feel we're looking at a development that runs from left to right. This applies not only to real lines but even to imaginary ones. An image of the evolution of man that is famous and erroneous in equal measure is called the March of Progress. It is derived from the ancient *scala naturae*, literally the stairway of nature but generally known as the Great Chain of Being, which represents a God-given, unchangeable hierarchical arrangement of the universe. It is a pecking order with God at the top, followed by angels and then human beings. Next come animals, below them are plants and finally, right at the bottom, simpler materials such as soil and rock. Combined with ideas about the development of life on earth that came into vogue after the publication of Darwin's *Origin of Species* in 1859, the idea grew up that evolution led from the simple to the increasingly complex, culminating in high-grade organisms with the near-divine human being as the pinnacle of development so far. That idea, which became established more widely after the publication in 1936 of a book by the American philosopher Arthur Oncken Lovejoy called *The Great Chain of Being*, has been depicted and reproduced thousands of times – not to speak of the endless stream of more or less comical variations on it. The great palaeontologist Stephen Jay Gould is said to have left a collection of several hundred versions when he died in 2002.

In its original form the March of Progress is a line procession of beings that, studied from left to right, are seen to be increasingly human. *Homo sapiens* takes the lead, self-consciously walking upright, while a hunchbacked species of primal ape lurches along at the rear. From a neutral perspective it's an absurd picture that makes no sense at all. Modern man, who chronologically brings up the rear, is at the front. He's followed by his own ancestor, who in turn marches ahead of his forebears. Yet we are untroubled by this, because our perspective is anything but neutral. We interpret the sequence as an imaginary line, a movement through time from left to right, so the image makes sense and everyone is in his proper place: the little hunchback comes first, then the others and finally, on the right, we ourselves. Moreover, each individual walks from left to right, in the direction of time, from his own preliminary stages towards his successor. This too fits with our perception of reality. In fact the picture accords not just with our experience but with our desires, since in the person of the

The March of Progress.

Guido Reni, *Atalanta and Hippomenes*, 1625.

man at the head of the line we walk, heads raised, towards the future. Were we to reverse the image it would seem as if evolution were moving in the wrong direction, as if we were walking further and further into the past, an ability reserved exclusively for Merlin, the wizard of Arthurian legend.

Our tendency to see pictures as naturally progressing from left to right is no recent invention. In 1625, for instance, Guido Reni made use of it in his painting of the contest between Atalanta and Hippomenes. Atalanta, the great huntress of Greek mythology, had hundreds of suitors on account of her many talents. To rid herself of the attentions of all those aspiring gentlemen, she promised to marry anyone who could beat her in a sprint. She knew what she was about, since no one could run faster than she could, but Hippomenes took up her challenge and managed to outdo her in a typically classical Greek manner, by cheating with a little divine assistance in the form of three golden apples. During the decisive race he dropped one of the apples every time Atalanta threatened to pass him. Each time he did so she was unable to resist the temptation to pick it up. Reni's painting shows one such fatal moment. Even without knowing the story, it's immediately clear what's happening: Hippomenes rushes forward, in the usual direction of movement, while

Atalanta turns back. Clearly she is doing something stupid. Indeed, she loses the race.

Comic strips offer even more opportunities for working with the symbolism of direction. W. A. Wagenaar, a psychologist at Leiden University and a Tintin fan, once tallied up all the acts and transitions in three Tintin books and discovered that in three out of four cases the characters move from left to right, even when their movement takes several frames to portray. More interestingly still, movements towards the left almost always have a bad outcome. The man who moves his finger from right to left to ring Tintin's doorbell falls unconscious on the doormat as soon as the alert reporter opens the door. When Captain Haddock crosses the frame from right to left in an attempt to escape, he is recaptured in no time. And so on. This is Tintin's Law, and it is reminiscent of the messenger in the theatre of the ancient world: entering on the right – and therefore moving towards the left – is a sign to the audience that something ominous is happening.

Similar rules apply to film. Cars, ships and planes mostly set off from left to right across the screen. If they move from right to left then we tend to be watching them arrive somewhere. Only in the case of great distances between generally familiar places does another criterion disrupt this pattern: geography. Ships travelling from Europe to America usually sail from right to left and vice versa. A train journey from Beijing to Moscow will be shown right to left, as will a plane trip from Calcutta to London. Western filmmakers comply with our habit of using maps with north at the top. They have little choice, since their overwhelmingly Western audiences see a ship sailing from left to right as moving from west to east.

Of course it's no coincidence that we interpret pictures the way we do. Experiments have shown that broadly speaking we actually look at them from left to right. The direction of movement in the picture coincides with the order in which we perceive it. Perhaps this is also part of the reason for the great popularity achieved and retained, through both libertarian and prudish periods, by one of the most remarkable paintings of all time, despite the fact that it is an image that makes the average American parent rush to place a hand over the innocent eyes of her – or indeed his – child. The painting is called *Gabrielle d'Estrées and One of her Sisters*, painted sometime around 1594 by an unknown artist referred to simply as belonging to the School of Fontainebleau.

It's one of a series of paintings of the same women in roughly the same setting, and in that sense it's reminiscent of a photo session by a modern-day photographer. This particular painting, however, is the most famous of the series by far.

We are looking at the woman with whom King Henri IV of France was deeply in love from 1591 until her death at the age of 28, but could not marry because he already had a wife. That did not prevent him from conceiving no fewer than three children with her and regularly conferring titles and their accompanying fortunes upon her. It is skilfully painted, with pleasingly self-assured mastery, especially when it comes to the main subject, Gabrielle herself. She is placed in the best position to emphasize her importance, where the viewer's gaze comes to rest after gliding over the painting from left to right. This also means that Gabrielle occupies the traditional position of the wife in double portraits, to the left of her partner, one result being that the light coming from the left as we look at her falls on her face, lighting it with agreeable softness. That's not all. The most important elements of the picture are on a fluid line running from top left to bottom right, from the face of Gabrielle's sister, the Duchess de Villars, to the hand with which she pinches Gabrielle's nipple between thumb and forefinger, and then on to the ring that Gabrielle is holding in a similar manner. The other lines of the composition harmonize with these so that they reinforce each other. All this makes the painting satisfying to look at, even a little slick.

If that were all then the picture would have excited few people other than admirers of apple-shaped breasts. The fact it has intrigued so many is attributable to two ways in which the canvas creates a sense of tension in the viewer. Firstly, of course, there is the bizarre pose with its incomprehensible symbolism: two women, one of whom affectedly holds the other's nipple between thumb and forefinger in precisely the way the other is holding her ring. It's unusual at the very least. The ladies are also startlingly naked, yet they sit or stand as if dressed to the nines. The second source of tension is less notice-able, and as far as we can tell it has in fact passed unnoticed; but perhaps precisely for that reason it contributes to the painting's mysterious power of attraction, the excitement that makes us look and go on looking: the composition's three figures are all left-handed. The duchess holds the nipple with her left hand, Gabrielle does the same with her ring, and to anyone who looks closely it becomes clear that even the seamstress in the background is embroidering with

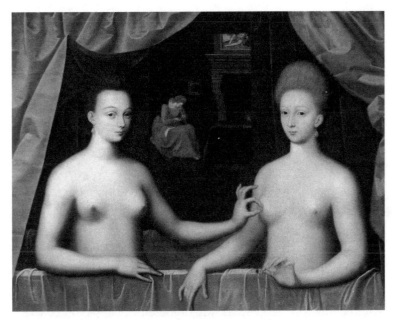

School of Fontainebleau, *Gabrielle d'Estrées and One of her Sisters*, c. 1594.

her left hand. This is unique, and slightly disturbing. It seems we subconsciously feel something is not quite right without being able to say what it is.

Studies show that we initially look at a composition from left to right, but why is this? It can't have anything to do with our eyes, since they move as easily in one direction as the other. So why not begin on the right, or somewhere in the middle? It seems that some people do.

In the days when Congo was a Belgian colony, in other words before 1960, mining engineers worried about the many mistakes made by poorly trained Congolese miners, and all the damage and accidents that resulted. They decided to do something about it. Wherever it was deemed necessary, clear, simple rules would be displayed so that there could no longer be any misunderstandings about the proper ways to use equipment or what to do in an emergency. This was not particularly easy, since workers recruited from the local population were almost entirely illiterate. Education would have been wasted, after all, on people who were to spend their lives smashing chunks of ore and rock out of the walls of subterranean vaults.

Nevertheless, a solution soon seemed to have been found. Instead of written instructions, simple illustrations, rather like cartoon strips,

The Long March, Shao Shan Mao Zedong Museum, Changsha.

would be used to show what was supposed to happen in concrete terms. Even the stupidest Congolese couldn't go wrong, they thought.

Not so. Instead of fewer accidents there were considerably more, and more mistakes were made too. After extensive investigation, someone eventually pointed out what was wrong: the cartoons were designed to be read from left to right. It hadn't occurred to anybody that although this might be obvious to a Belgian, who had been to school and had grown up in a completely literate world, it was not at all clear to an illiterate Congolese, who had no idea where to begin. As a result the local miners regularly drew quite different conclusions from those intended.

Indeed it does turn out that the kind of rules of looking that we are discussing here depend to a great degree on the direction of writing. Wagenaar's Tintin's Law works the other way around with right-to-left writers such as Israelis and Arabs, as any cartoon strip created by them demonstrates. China has never had a true left-to-right tradition. Chinese characters were traditionally written from top to bottom in columns running from right to left, so the Chinese look at images differently from the way we do. In the Mao Zedong Museum in Changsha, for example, a painting is on show that depicts the Long March. It's a fantastic piece of in-your-face propaganda, with a mass of strong, resolute fighters on their way to inevitable victory under the inspiring leadership of Mao Zedong. Or is it? To us it seems a strange image, since the entire stout-hearted company is marching ominously towards the left of the frame, in other words in the wrong direction. Such a picture would never be composed that way either in Russia or in the Western world.

Reading and writing, the skills that have done more than anything to make the tempestuous cultural and economic development of humanity over the past six to seven thousand years possible, turn out to influence us even in contexts that involve no words at all.

Rembrandt van Rijn, *Abraham and Isaac*, 1634.

18

Dead Men and Voluptuous Women

Some time between 1655 and 1658, Nicolaes Maes painted the sacrifice of Isaac. The patriarch Abraham is on the point of taking the life of his son Isaac for the greater honour and glory of the Lord when an angel restrains him at the very last moment. God has already seen enough to be convinced. It's an extremely familiar subject, but Maes, supreme craftsman and artist that he was, has done his best to turn it into something extraordinary. The result is an uneasy, unstable picture, as becomes all the more clear when we compare it to the standard composition Rembrandt used some twenty years earlier to depict the same scene. Rembrandt's painting somehow makes more sense. It seems more natural. This again probably has to do with the way we read.

Just as we see the sloping lines in graphs rise or fall, paintings have a rising and a falling diagonal. The rising line begins at the bottom left, the descending line at the top left. They are of crucial importance, as demonstrated by the fact that they constitute the main difference between the sacrifice scenes painted by Maes and by Rembrandt. In Rembrandt's painting, Abraham's knife has threatened until a fraction of a second before to move down the falling diagonal and cut Isaac's throat, but in Maes's picture, should Abraham raise his hand the knife will meet poor Isaac's throat by moving diagonally upwards, from right to left. This is highly unusual. Murder attempts in paintings usually happen from left to right along the falling diagonal, as in Rembrandt's depiction. The weapon, whether a knife, sword, axe or club, usually strikes the victim from high on the left, and in most cases the victim too is aligned with the falling diagonal. He either lies along it, wounded or dead, or flees towards the bottom right.

An outstanding example of this use of the diagonal is Francisco Goya's *Highwaymen Attacking a Coach*. It's a dramatic scene, one that

Nicolaes Maes, *The Sacrifice of Isaac*, c. 1657.

suggests there was nothing new about the cowboy films made many years later, but note the positions of the various characters. The leader of the highwaymen stands high on the coach-box, triumphant, pointing his gun downwards with studied nonchalance as he gazes along the descending diagonal at the distressed passengers. His accomplices, especially the one on the left, are busy dealing with a traveller who is still putting up resistance. He's about to use his knife to silence the poor man for ever, and that knife too will strike home along the falling diagonal. Two women beg for their lives, on their knees, hands raised. The one whose face is visible looks back up along the falling diagonal at the highwayman on the left, who has his eyes fixed on her. The male passengers are dead, or soon will be, and all three are lying along the descending diagonal. Note too that Goya adheres to Tintin's Law: on its doomed journey, the coach travels from right to left across the canvas.

It's not particularly odd, when we stop to think about it, that blows and stabs take place along the falling diagonal. For a start, most thrusts of this kind are in a downward direction. Second, victims have a tendency to fall over and artists often anticipate this for dramatic reasons, so the murderer or would-be murderer almost automatically stands above the losing party. The preference for left-to-right thrashing and stabbing is explicable too, since for Westerners that is the standard direction in which such things are depicted. The murder weapon gains as it were extra speed and force when it moves from left to right. The same goes for flight. He who flees must be quick and the suggestion of speed is not helped by having him scrabble against the most natural direction of movement. Scenes of Adam and Even being expelled from paradise therefore almost always have the same composition: the Angel of Vengeance floats somewhere in the top left-hand corner, while Adam and Eve leave the Garden of Eden at the bottom right.

More striking still is the fact that the victims of lethal violence often lie along the falling diagonal, even in photographs. Take press pictures of mafia victims on village squares in southern Italy, which appear in the newspapers from time to time. They lie motionless, naturally, so the photographer can take his time choosing the angle that suits him best, which is very often one in which the body lies head first down the descending diagonal, precisely as in Manet's *Dead Matador*. That painting in turn closely resembles the seventeenth-century *Dead Soldier*

Francisco Goya, *Highwaymen Attacking a Coach*, 1786–7.

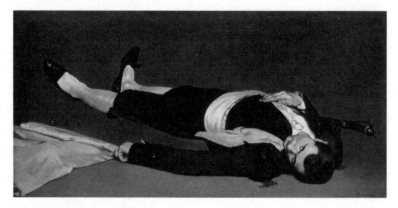

Edouard Manet, *Dead Matador*, 1865.

by an unknown Italian painter, currently in the National Gallery in London, which some people claim had a strong influence on Manet, although Manet experts contest this. The experts may be right, since victims of violence posed in this way have simply been placed in the standard position.

The reason why the falling diagonal is so popular in these pictures of corpses cannot be a direct consequence of the way we perceive movement in pictures. After all, the dead do not move, so in paintings they shouldn't give any impression of dynamism. Is it perhaps the descending aspect that's relevant here? Could it be a way of indicating that the person depicted came to a bad end? Does the body almost literally slide out of the painting towards doom, dispatched by violence that apparently originates at the top left?

This is not such a ridiculous idea as it might seem. In Goya's painting the power relationships coincide with the course of the violence, from top left to bottom right. The chief highwayman and his main accomplice look down with superior expressions at the pleading woman at the bottom right, who looks up at them submissively in the opposite direction. This slipping away downwards towards the right emerges even more clearly in Dirck van Baburen's painting *Prometheus Being Chained by Vulcan* (1623), which can be seen in the Rijksmuseum in Amsterdam. Prometheus is lying in the worst possible position, chained up in the lower right-hand corner, against the frame. It seems as if but for that frame he would slide off the canvas completely. Like Manet's matador and all those other prostrate victims, he lies head first, unable to see what he's sliding towards. It's a picture of utter powerlessness.

A victim of the Mafia photographed in 1992 according to the rules of visual art. Note the many falling lines.

Anti-Mafia demonstration in Naples, 2005.

Dirck van Baburen, *Prometheus Being Chained by Vulcan*, 1623.

Most paintings of murders, assaults and torture adhere to the same pattern. Power, especially oppressive power, comes not only, as Mao Zedong would have it, from the barrel of a gun but in most cases from the top left, moving with gravity and in the direction in which we read.

Of course other considerations apply as well. Compositional difficulties or competing conventions and the desire to treat a subject in an entirely new way can lead to an image that appears to go against the grain. Nicolaes Maes's *The Sacrifice of Isaac* is one example. But often the artist nevertheless submits to these laws of perception, perhaps without even being aware of it. Take for example Goya's most famous painting, *The Third of May 1808*, which depicts the execution of 43 rebels in Madrid, one of a number of harsh reprisals by Napoleon's occupying armies in Spain for the Madrid uprising against the French.

The firing squad carries out its bloody work from right to left, giving Goya the opportunity to portray the soldiers from behind as an anonymous killing machine. At the same time he allows us to look straight into the emotional faces of the victims, who meet their tragic fate no less conventionally along the falling diagonal, as further emphasized by the angle of the hillside in the background. Goya has the next group of condemned men approach from the distance on the right. They are moving leftwards against the falling diagonal, trudging up towards the place of execution, high on a hill.

It's very different for women. Not many women die in paintings, aside from the occasional Lucretia or Cleopatra. They far more often lie naked in charming postures, draped across chairs and settees, or slumber quasi-innocently in delightful groves. At best they have a good time with a swan, like Leda, or in the case of Europa lovingly embrace a tempting bull. But does this mean that the effect of the diagonal direction, so powerfully present in scenes of violence, is irrelevant in portraits of women?

Not when it comes to violence or unwilling submission, which often means sexual violence. When Gustav Klimt depicted Leda yielding to

Francisco Goya, *The Third of May 1808*, 1814.

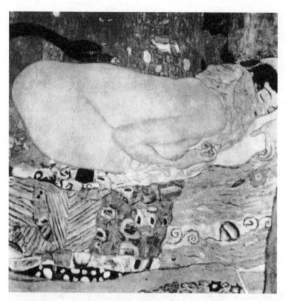

Gustav Klimt, *Leda and the Swan*, 1917. The painting was lost in 1945 when German troops set fire to Immendorf Palace.

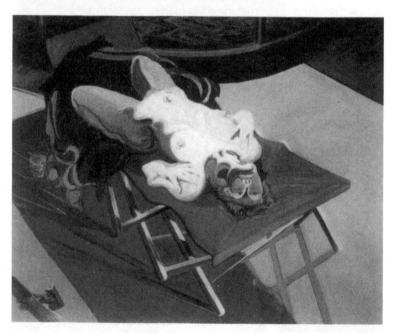

Norbert Tadeusz (1940–2011), *Scorpion*.

Titian, *Venus of Urbino*, 1538.

the swan, in a painting tragically lost to fire in 1945, he showed her on the point of being penetrated from left to right. In the strange painting *Scorpion* by Norbert Tadeusz there's a particularly powerful suggestion of subjugation and sexual assault. On the face of it nothing is happening, but the sunbathing woman looks even more tormented than Baburen's Prometheus. Here too the victim lies head first along the falling diagonal.

These compositions are fairly exceptional, however. Most female nudes appear to be in fine fettle. Nevertheless, the diagonal is of importance in their pictures too, if rather less prominently so.

The best example of a recumbent nude is perhaps Titian's *Venus of Urbino*, a painting that was a source of inspiration to dozens of later artists. The *Olympia* that helped make Manet famous is directly derived from it. Despite the fact that Venus is lying along the falling diagonal, the painting is not ominous in any way, if anything quite the opposite. Titian's beauty resolutely looks us in the eye, clearly the mistress of the situation, and partly for this reason she seems a little naughty. The lap-dog at the foot of the bed – such a powerful symbol that even today many prostitutes go around in the company of a similar animal – is an additional indication of her boldness.

Impertinence seems to be a hallmark of nudes depicted along the descending diagonal. They often seem rather less virtuous than average, making eye contact with the viewer more often and with a slightly cheekier look. This is almost inevitable, since to make the light, which usually shines from the left, fall softly on her face, the artist has to turn the woman's head slightly towards us. This suggests she's deliberately looking at us. Nudes that lie along the rising diagonal more often sleep the sleep of the virtuous, or imagine themselves unobserved. It's as if they are protected from sliding towards moral turpitude by looking from the right side of the painting towards the left. This outcome is far less obvious than the effects we've noted in scenes of violence, and there are far more exceptions to the rule, but it's difficult to avoid the impression that for female nudes too, the falling diagonal is a slippery slope – the difference being that they find themselves not in a physical danger zone but in a moral one.

19

Mary's Little Troublemaker and Other Portraits

Anyone who has regularly examined depictions of the Madonna with child in galleries and museums must surely have wondered why painters so rarely manage to portray an agreeable-looking divine infant. Instead of a podgy little pink delight that sends everyone into ecstasies, we are generally treated to a horrid, blubbery, prematurely aged dwarf. There may well be entirely prosaic, practical reasons for this. Real infants are as intractable as they are endearing. They refuse to sit still, cry at the most inappropriate moments and need almost constant care. In short, they're incapable of posing properly. A great many painters must therefore have done most of the job from memory or with the aid of a doll.

What those painters probably didn't know is that they were responsible at least in part for their subjects' objectionable behaviour, at least if we're to believe the research published by Canadian child psychologist Lee Salk from 1960 onwards. Salk investigated how mothers prefer to carry their offspring and discovered that over 83 per cent of right-handed women in his study held the child with their left arm. This appears logical, since it leaves the right hand free for other tasks, but there seemed to be something else going on, since even of the left-handed mothers Salk studied, almost 80 per cent preferred to use the left arm to carry an infant.

That preference, Salk decided, is the result of what he called 'imprinting'. In the womb the child had become used to hearing and feeling its mother's heartbeat and after birth that same auditory input had a calming effect. He played the sound of a normal heartbeat to 100 infants in a hospital maternity ward, at the volume they would have heard it before birth, and compared them to a control group. Sure enough, the first group turned out to cry less, to go to sleep more easily and even to grow more quickly than the children deprived of heart-music. It's true that our hearts are only a fraction to the left of the centre of the chest cavity, but because of the greater pressure in the left ventricle and therefore in the left breast, the pumping of the heart sounds louder on that side. If you hold your child on the left, Salk concluded, it will make less noise. Which is exactly what mothers like to hear.

Painters generally don't know a great deal about baby-care. Their concern lies with the composition of the painting, which doesn't always coincide with the interests of mothers and children. This was demonstrated in 1973 by Richard Uhrbrock, an American psychology professor with a great interest in Madonnas. Of the more than 1,100 Madonnas he examined, 45 per cent held the Christ child in the left arm, 38 per cent in the right. The remaining infants, some 8 per cent, sat in the middle of the lap. Uhrbrock felt the percentage of left-arm-sitters was remarkably high, but if Salk is right it's actually remarkably low. It means that in at least 20 per cent of their sittings the painters of the pictures examined by Uhrbrock must have personally ordered the model to place the child on her right arm. Infants are creatures of habit, whose behaviour is unlikely to improve when they're made to comply with unfamiliar whims, so there must have been good reasons for all those painters and models to have insisted on their choice in the face of the inevitable grouching and whining.

One of those reasons is demonstrated by the famous painting *Las Meninas*, or *The Maids of Honour*, by the Spanish painter Velázquez. It's an artfully arranged composition that nevertheless manages to look like a domestic snapshot of part of the personal entourage of King Philip IV of Spain. The subject is ostensibly the little blonde princess in the splendid white outfit at the centre. She is Philip's only surviving child, the *infanta* Marguerita, and she is accompanied by her two maids of honour, Maria Augustina Sarmiento, who holds her right hand, and Isabel de Velasco. But there is a great deal else of interest in the painting, including Velázquez himself, who stands to the left of the canvas, busily painting a large picture, so large in fact that his canvas has been plumped right down on the ground. It's as if we, the viewers, are ourselves the subject of his next painting, although, as we will shortly see, we are not.

Clearly the painter is right-handed. If we imagine that the *infanta* and her entourage were not in the picture, then Velázquez is using the most simple standard composition for a portrait by a right-handed artist. The easel stands in front of the painter slightly to his right, the

Diego Velázquez, *Las Meninas*, 1656.

subject slightly to his left, in the same direction as the viewer. As we can tell from the patch of light on his forehead, the light is shining from the painter's left, so that the unfinished painting will be well lit, but this also means the light is falling from his left onto the subject.

Las Meninas is therefore painted from the position of the model for this fictional canvas, so from our perspective the light falls from right to left, making the work an exception to the rule. In traditional portraiture the light shining across the canvas is far more likely to fall from left to right, exactly as Velázquez shows within the painting. In collections held by the Rijksmuseum in Amsterdam and the Mauritshuis in The Hague, for example, this is true in more than 80 per cent of cases. The fall of light has consequences for the position of the person or persons in the picture. If they turn slightly to the right (from our perspective), away from the source of light, then their faces remain in shadow. If they look towards the left of the picture instead, their faces will be lit up. That effect too can be seen in *Las Meninas*, but in reverse. The *infanta* looks in the direction of the beams of light and her face is not only well lit, it's clearly a point of special interest in the picture. She is the most important figure among the group of girls, more important than Isabel de Velasco, whose face is in shadow since she is looking away from the source of light, and more important than the dwarf María Bárbola who, even though she is closer to the window, is looking almost directly at us and therefore catches the light rather less.

In the standard composition, therefore, the subject of the portrait looks towards the left of the picture, showing the painter and viewer his or her left cheek. This is one of the reasons why Madonnas are twice as likely to hold a child on their right arm than we would expect from Salk's data. A Madonna and child is a composition that can easily be painted in a studio. After all, the model is usually an anonymous girl, not a demanding client who must be catered to in her own home. The studio would be laid out for maximum efficiency, which for a right-handed

Two classic versions of the Madonna and Child. On the left Leonardo's *Madonna Litta*, 1490, on the right Albrecht Dürer's *Madonna of the Pear*, 1512. The light falls from the left, shining directly on Mary's face and softening its contours. As usual, the babies are fairly hideous.

artist meant the light would fall from his left. To catch the full light without any additional artifice, the model would therefore have to turn her
head to the left as we look at her, in other words to her right, but at the
same time a Madonna needs to look down lovingly at her child, which
must therefore sit at her right breast. Were it on her left side, she would
be looking away from the infant, as if she'd rather be without it, which
of course would not do at all.

The angle of the light, however important, is only one factor in determining the appearance of a portrait, which is why by and large there are
just as many portraits in which the subject shows the right cheek as there
are in which the left cheek is turned to the viewer, although according
to some the left cheek is slightly overrepresented. If we treat portraits of
men and women separately, however, the proportions change dramatically. In the Rijksmuseum and Mauritshuis collections, almost two out
of every three men show their right cheek while for women precisely the
reverse applies. Other collections produce similar figures. Clearly this is
something we need to explore further.

   Quite a few explanations have been put forward for the suspected
slight preponderance across the board of left-cheeked portraits. The

simplest is that it's more natural and easier for a right-handed person to draw a profile facing towards the left of the picture. This makes sense if we consider spontaneous sketches made without a model, or mindless doodling during boring meetings, but it would say little for the craftsmanship of a professional painter if he allowed himself to be guided by such considerations. Other theories refer to the social relationship between the artist and the model, or even to the way our brains recognize faces. None of them are satisfactory, for two reasons.

First they assume that all things being equal there is an absolute preference. In other words if a model of a given status prefers to be shown from a particular direction, then the same must apply to all models of the same status. Similarly, if it has to do with the way the brain recognizes faces, then the same preference would be at work in all cases. But coins show that this is not so. They always feature kings, emperors, gods or other superior folk and we look at them every day with the same brains, yet there is no way of guessing which profile will be used. The same Uhrbrock who studied the baby-carrying arm discovered that on American coins and commemorative medals the person portrayed looks to the left in two out of every three cases, but in the large collection of European coins at the Hamburger Kunsthalle, which covers 25 centuries, the proportions are almost exactly the reverse. Moreover, none of the theories explain why the difference in orientation between portraits of men and of women is so marked.

As to the latter question, *Las Meninas* may once again provide an answer, since not we but the royal couple, Philip IV and Mariana of Austria, are the models for the portrait that Velázquez is shown painting. They can be seen in the mirror in the middle of the back wall of the room.* Their mirror image shows that Mariana must be standing to the left of her husband, and this cannot be coincidental. For a double portrait, certainly until the eighteenth century, this was the standard arrangement. Since man and wife invariably stand or sit turned towards each other, such portraits almost always show the left cheek of the woman and the right cheek of the man.

---

* Some experts are of the opinion that the royal couple are not the subject of Velázquez's attention, since no double portrait of them is known and because the canvas shown in the picture is too large for a portrait. But in that case what are they doing at the spot where the model should be? Furthermore, what Velázquez professes to show, *Las Meninas*, cannot possibly have been painted from life. The fact that he was not in reality working on a double portrait therefore tells us nothing.

Albrecht Dürer, double portrait on separate panels of Hans and Felicitas Tucher, 1499. The light falls from the front left, emphasizing the angular face of the self-assured Hans and the soft, round features of his rather sceptical-looking wife.

We find plenty of examples of the same type of composition in double portraits on two separate canvases. There are almost no couples in portraits in the Rijksmuseum with the man standing to the left of his wife, and none at all among those painted prior to 1700. It is a traditional arrangement that shows we are dealing with a respectable married couple, and one that still applies in the rules for formal church weddings; once the ceremony is over, the bride leaves the church on the left side of her new husband. This tradition has apparently had a profound influence not only on the double portrait but, via the double portrait on separate canvases, on the composition of portraits of lone individuals as well.

The strong tradition of portraying women with their left cheeks to the fore and men the other way around therefore seems to be derived from rituals and etiquette, but there is probably more to it than that. The phenomenon is reinforced by the influence of the angle of the light on the composition. Even in right-cheeked portraits the light generally falls from the left, so the model is looking away from the light to a greater or lesser degree. This does not necessarily mean that the face is obscured, but it does tend to emphasize the edges of the profile, strengthening the curve of the nose and the lines of the chin and forehead. This kind of lighting tends to stress sternness, willpower and other characteristics traditionally attributed to the male. The softness, roundness and elegance that are thought of as typically feminine are not aided by such angular lighting; they are more apparent when the light bathes the whole face equally, in other words when the face is turned towards the left of the picture, where the light is coming from.

So it came about that a great many posing Madonnas, in order to display their femininity and their soft features, were forced to treat their children with unpleasant severity.

20

Little Johnny Cries to the Left, Little Johnny Smiles to the Left

Human faces are remarkable things. Everyone knows that the left and right halves may differ considerably. Like the rest of our bodies, faces are only broadly symmetrical, a fact that is not without consequences. Symmetrical faces are seen as a hallmark of beauty. There's an assumption that we can explain this in evolutionary terms. Regular features are said to indicate good health and a flawless package of genes, and people with symmetrical faces are therefore valuable partners in the struggle to reproduce.

Less well known is the fact that the two halves of the face have different roles, both in the recognition of someone's identity and in the interpretation of his or her emotional state.

Generally speaking people can effortlessly express at least six different emotions on their faces: happiness, sadness, surprise, fear, disgust and anger. Of course we can convey more than that with our facial expressions, but these six occur in all cultures and are equally easy to produce and to recognize in every population group. Oddly enough it's the left side of the face that does most to determine which emotions others see in it. This was made clear many years ago when research subjects were shown two pictures of a face, each the mirror image of the other and in all other respects identical, in which one side was cheerful while the other looked glum. Around eight out of ten people turned out to base their assessment of the depicted character's mood mainly on the left side. If the left half of the figure's face looked unhappy, people thought he was unhappy, if only the right half looked sad, he was seen as rather more cheerful.

When it comes to the straightforward recognition of a person based on his or her face, the right side weighs more heavily. Tests show that people feel an image of the right half of a face filled in with a mirror

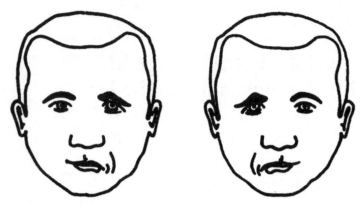

Eight out of ten people feel that the face on the left looks more sorrowful than the face on the right. In fact they are identical, except that one is the mirror image of the other.

image of that same half bears a greater resemblance to the person depicted than a composite photo made of two images of the left half. In some sense we resemble the right side of our faces more than we resemble the left.

This phenomenon probably has to do with the difference between the two halves of our brains, which each have their own specialist functions. The left half of the brain is good at everything that relies upon calculation: counting, arithmetic, logical reasoning and much that has to do with language. Among the things the right cerebral hemisphere concentrates on are the interpretation, recognition and recollection of images. Of course there are a great many things that the two halves of the brain do together, if only because many complex tasks make use of simple functions, some of them located in one half of the brain and some in the other. That cooperation between the two hemispheres is made possible by connections between them at various levels, known as commissures, which work like bundles of telephone lines in a large office complex. The biggest by far is the *corpus callosum*, a broad band of tissue that is unique in making direct contact with both halves of the cerebrum.

In the case of our eyes, which in a sense are bulges in the brain that stick out through the skull, information coming from the left half of each retina is initially processed by the left half of the brain, and vice versa. So if we look directly at a face, its left half, which is on the right side of our visual field, is registered on the left half of the retina and that half is therefore initially processed by the left cerebral hemisphere. The right side of the face, to the left as we look at it, is

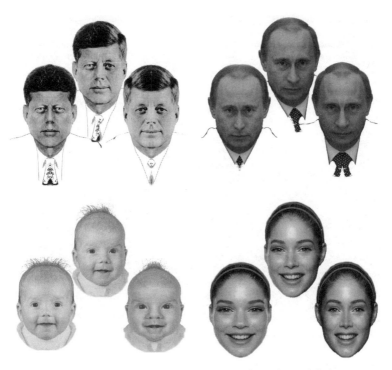

Four famous people and their half-faces. Clockwise from the top left the American president John F. Kennedy, Russian ruler Vladimir Putin, supermodel Doutzen Kroes and baby Cleo. In each case the middle picture shows their real face, with below it a doubled left half on the right and a doubled right half on the left. Even Kroes's extremely symmetrical face looks more like its right half.

registered on the right half of the retina, which sends signals to the right half of the brain.

It therefore seems logical that we base our recognition of faces on their right side, since that information goes straight to the right half of the brain where our capacity to recognize faces is located. Information about the other side can reach that part of the brain only via a detour through the *corpus callosum*. This works fine too; we can instantly recognize a face even if its owner covers the right half, although it does take us a tiny bit more effort. Why is it then that in recognizing emotions on faces, a fairly subtle piece of interpretation based on visual information alone, we mainly take account of the left side of the face?

One explanation that has been advanced is that there is simply more to be seen on the left side of a face, since it's controlled by the right side of the brain, the side that is more emotionally orientated. But

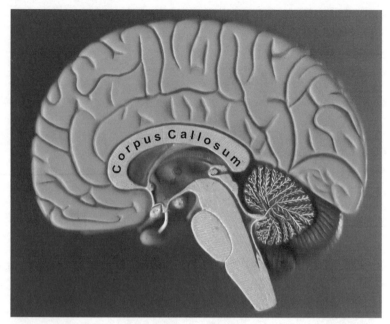

Cross-section of the brain in between the two cerebral hemispheres. We are looking at the right side, with the right half of the bisected *corpus callosum* at the centre. The white brain stem below it continues downwards as the spinal column.

even if this is correct it does not solve the problem. It's certainly not the case that emotions are exclusively the prerogative of the right side of the brain and even if they were, the advantage of the greater expressiveness of the left side of the face would be cancelled out by the emotional hamfistedness of the left half of the brain, which we use to observe that part of a face. The following explanation may therefore be more convincing, although it too is based on an unproven assumption.

Although emotions are not entirely the preserve of the right cerebral hemisphere, it does seem true to say that the right brain is more emotionally oriented than the left. It could therefore be the case that the left half of a face, controlled as it is by the right side of the brain, is more expressive of emotions. If so, then the right side must be less changeable. This is precisely what makes it a better guide to simple recognition: in all circumstances it looks more like itself than the left side, which has a greater tendency to change according to the mood of its owner. As an extra bonus, when we look at a person the relatively unchanging half of his or her face is connected by the shortest possible route to the portrait gallery in the brain that enables us immediately to recognize family, friends and acquaintances, which is also located in the right brain.

The left brain is better suited to the recognition of the various expressions of emotion on a face. The recognition capacity of the right brain concentrates on making a perfect match between what we see and an image stored in the memory. Fleeting variations only make that task more difficult, whereas subtle changes suit the left brain rather well. Once our right brain has determined whose face we are looking at, our left brain can compare that image with what it sees on the left side of that same face. It then needs to make a complicated subtraction sum: what the left brain sees minus the right-brain-determined standard face gives the emotion that is being expressed. This information can be elaborated further, enabling us to work out exactly what emotion we are witnessing. Although this seems to involve a lot of traffic between the two halves of the brain, each half does exactly what it's best at: the right brain is responsible for the recognition of a face and the interpretation of emotions at a high level, while the left brain makes calculations at which the right brain is less adept. In this way optimal use is made of the information we perceive with our eyes.

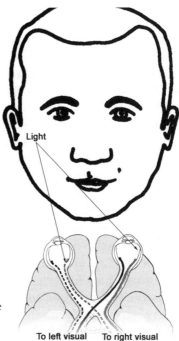

When we look directly at someone, the light from the right half of their face arrives on the right side of our retina. We therefore use the right half of the brain to 'see' the right half of a face.

21

The Circle Dance of the Alphabet

In April 1949 a remarkable photograph cropped up in various places around the world, showing a group of Yemeni Jews in a reception camp near the seaport of Aden. They are on their way to Israel and they're all crowding around a Torah. One has the book in front of him in such a way that he can read it in the normal Hebrew manner from right to left, in lines that run from top to bottom. A second is sitting off to the left of the Torah and is therefore forced to read columns of text that run from top to bottom and from left to right. In the foreground another man is reading the text upside down and the rest too, from various angles, are doing their best to look at the pages.

It's difficult for people in the rich world to imagine, but clearly these gentlemen are at ease with their unconventional reading positions. A scarcity of books, such that one copy had to be shared between three or four schoolchildren, had caused them to learn to read from various angles. Why not, in fact? There's no law of nature that says, for example, that our letter A must stand with two feet on the ground; in fact, there was once a time when it didn't. Originally, in the Phoenician alphabet, it was upside down, forming a pictogram of an ox's head with horns. Later it came to lie on its side and only when the Greeks adopted it did the two 'horns' come to rest on the ground.

We may wonder how the men in the photograph wrote, assuming they had learned to do so. Did they orientate themselves in the same way as for reading, or did they write in the standard Hebrew manner, in horizontal lines from right to left? Did this affect how well they could write? The profound effect that the direction of our writing has on our observation of reality has led many people to conclude that, far from being a matter of chance, it's determined by nature, a form of behaviour that somehow flows from our genetic inheritance, probably

Different
perspectives
on a copy of
the Torah.

connected in some way to the predominance of right-handedness. It's generally assumed that right-handed people naturally prefer to write from left to right, left-handers in the other direction.

This reasoning, to which many adhere, teachers in particular, suggests that ours is a fundamentally right-handed script. Arguments have been put forward in support of this claim. It is said to be more natural to use a right hand to write from left to right because it will then move away from the body. This conviction is in turn based on the assumption that movements away from the body are more natural than movements towards the body, although it's far from clear why this should be the case. A second argument often advanced is that right-handers who write from left to right pull the pen across the paper, whereas if they write from right to left they are required to push it, which inevitably leads to smudges, and to bent and split nibs.

Anyone who replies that people elsewhere in the world are as predominantly right-handed as we are but for centuries have written from right to left to the full satisfaction of all concerned are told this is merely an illusion. The example given is almost always that of Hebrew. In that language, although the letters are written one after the other from right to left, each letter is formed from left to right. This is also the reason, people go on to say, why Hebrew is still written in block letters and does not have a flowing, joined-up form. Sometimes Chinese is brought to bear as well, a language traditionally written from top to bottom. The Chinese too, after all, form each character from left to right. The rules for forming letters and characters in these languages are said to indicate that basic biology will always show through, even in the case of right-to-left handwriting: people are naturally inclined to work from left to right.

We will return later to the push and pull movements performed by left- and right-handed people, and to those bent nibs, inkblots and smudges, but this last assertion – that even if the script runs from right to left or top to bottom the individual letters or characters are formed from left to right – is undeniably true in the case of Hebrew and Chinese. Yet it does not at all follow that there's anything natural about writing from left to right, or even that letters and symbols are always formed in that direction. To see this we only need to step across to Israel's neighbours, the Arab countries. Arabic is written in the same direction as Hebrew, but it's a joined-up script with letters written in a fluid motion from right to left. Block letters don't exist in Arabic. Of course Arabs are no less right-handed than we are, in fact in Arab cultures the left hand is subject to a far stricter taboo than in the West. As a young Arab you can forget about trying to write with your left hand. It's simply not tolerated.

Arabic is a rather unusual form of writing, a script with a hugely important calligraphic tradition that has its origins to a great degree in the Islamic ban on the representation of the human figure, a rule that in earlier times was strictly enforced. Calligraphy was one of the ways of creating visual art and decoration within those limitations, and a

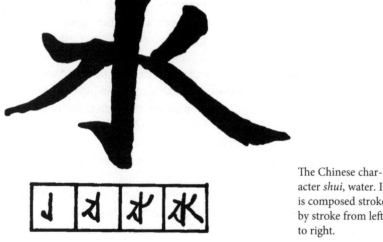

The Chinese character *shui*, water. It is composed stroke by stroke from left to right.

wide range of decorative forms of handwriting were developed to that end, in which the priority lay with beauty rather than readability. Of course decorative writing exists in Latin script too, as seen in those curlicue letters we like to use on charters and diplomas, but the practice was taken a good deal further in the Arab world.

Most forms of everyday Arabic writing stand out from other types of script in their elegant, flowing style, and they are usually written not on the line like Latin script but around the line. Each word dances from just above the line on the right to just below it on the left, like an endless series of waves meeting the coast. The shape of many letters varies significantly depending on the place they occupy between neighbouring letters.

Arabic script was impossible to produce satisfactorily on a typewriter, since it could not be reduced to a straight line of unchanging letters. Generally speaking this also applied to the typesetting machines used in the printing of works in Arabic. It was a limitation that had all kinds of deleterious effects on the Arab world, which the Enlightenment passed by and which in any case did not lay great emphasis on reading and writing. Not until about 1990 did the computer start to overcome the problem. Modern word processors and computerized typesetting systems can produce perfectly respectable Arabic script.

Would it not be ironic if Arabic, of all languages, had been written for many centuries contrary to the 'natural writing direction' – a script so much geared to handwriting that it can be achieved mechanically

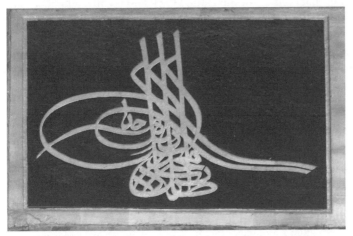

Monogram of the Turkish sultan Mehmet II, a masterpiece of calligraphy from 1223 that pays little heed to legibility and is all the more beautiful as a result.

only by advanced computer programmes? If that were the case, could it ever have spread through large parts of Asia, from Turkey to Indonesia, and across all of North Africa and most of East Africa? Could it have maintained its hold so effortlessly right up to the present day in the entire Arabic-speaking world from Morocco to Iraq, in Iran and in Pakistan, without ever being replaced by a less 'unnatural' alternative? It seems highly unlikely.

The improbability becomes even greater when we consider that our alphabet and Arabic script both emerged in the distant past from the same northern Semitic tribe to which the Greeks belonged. The Greeks, who with their block letters presumably had far less difficulty with their supposedly unnatural direction of writing, began to write from left to right in about 800 BC. Three centuries later that became their sole norm. So why would the choice have been made to write joined-up Arabic in the 'wrong' direction?

In reality Arabic is not awkward to write at all. Its dancing, artistic character arises from an optimal adjustment to the requirements of writing from right to left with the right hand. If we take a close look at the dance-steps that Arabic writing makes, then we see that each word is written on a slight diagonal from top right to bottom left. Each new word starts well above the line. Given that the paper is perpendicular to the body, this also prevents the right hand from brushing across the letters it has just produced and smudging them.

The fact that the two most widely used types of script, Arabic and the group that consists of the Latin, Greek and Cyrillic alphabets, are written opposite ways around is far from the only evidence that there's no such thing as a natural writing direction. Many different ways of writing have been used in the past. Until the sixth century BC some Greek inscriptions were written boustrophedon style, most commonly those in either Etruscan or Demotic, a script that in Egypt developed

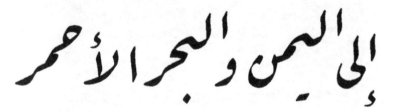

An example of Ruq'a, an Arabic script in daily use in the Middle East that is excellently suited to right-handed writing. Each word is written from right to left on a falling line.

from hieroglyphics. Boustrophedon means roughly 'turning the ox' in classical Greek and refers to a way of writing in which the lines were written alternately from left to right and from right to left, the way a farmer ploughs a field. Sometimes individual letters were reversed as well, sometimes not.

Boustrophedon is not at all illogical. Instead of moving the hand and the focus of the eye all the way from one edge of the writing surface to the other every time you start a new line, you read and write from the point where the previous line ended. Since the system never spread much further and soon disappeared from Greek, we can assume that this advantage did not outweigh the disadvantage of dealing with two reading directions at the same time. This probably has to do with the fact that in boustrophedon script the word-images – and in some cases even the letter-images – are not constant. The same word can appear in mirror image depending which line it is on. The word Mirror can suddenly appear as rorriM or even ɿoɿɿiM. If we spell out the word letter by letter, this is not much of a problem, but that's simply not the way we read. Once we've learnt to read quickly and fluently we recognize entire words at a glance, which in a boustrophedon text is at least twice as difficult.

A way of writing that attempts to combine the advantages of both boustrophedonic and modern ways of writing has been found on Easter Island – perhaps the unlikeliest place on earth. In 1868 a German missionary called Zumbohm found there a set of *kohau rongo rongo*, or message boards, pieces of wood inscribed with what is probably an ideogrammatic script (like hieroglyphics). The remarkable thing about them is that they are written in a special kind of boustrophedon script in which each line is turned on its head in relation to the one before. This made it easy to continue writing from the place where you had to

Easter Island script found on a message board. Not only is it written boustrophedon style, each line is upside down in relation to the previous one. The entire piece of wood would have to be turned through 180 degrees at the end of each line.

stop, while at the same time ensuring that all words were written in the same direction. The disadvantage, of course, was that at the end of each line the whole plank had to be turned around, which cannot have been terribly convenient.

Unfortunately, when Zumbohm made his discovery there was no longer anyone living who could read the script, so the meaning of what is written on those little planks of wood remains a mystery to this day. What the Easter Islanders did think they knew was where the script came from. According to traditional stories it was brought to them by their first king, Hotu-Matua, who arrived on the island by boat in the twelfth century. Whether or not this is true we cannot be certain, but striking similarities have been found between some of the symbols on the *kohau rongo rongo* and characters of the Indus script once used in parts of the Indian subcontinent.

The Chinese do things differently again. They write their characters in columns that run from top to bottom and are placed next to each other to be read from right to left, while each individual character is formed from left to right. But there have also been scripts that ran vertically downwards with columns arranged left to right, the ancient Mongolian script for example, or that were written as a vertical boustrophedon.

In truth there is no imaginable direction that has not been used by some linguistic group or other. Although people have been known to change the direction of their handwriting, think up an entirely different system, or adopt the style of their neighbours, until the sixteenth century no clear shift towards a particular direction of writing can be discerned. It seems a majority agreed early on that it's more convenient to work from top to bottom than from bottom to top, but there's no trace of a consensus when it comes to the horizontal direction, except that boustrophedon systems never really caught on. One extremely turbulent region in this respect is present-day Turkey, where the script was originally boustrophedon Hittite, after which people switched to writing Lydian, which ran from right to left, a language superseded by Greek, which is written in the opposite direction, before the Ottomans introduced the right–left Arabic script from 1453 onwards in which their Turkish language was written until 1928 when, under Atatürk, it was given a modern Latin alphabet of its own that ran from left to right once more.

However often and however categorically it is posited, the idea that writing from left to right is more natural seems to be primarily a

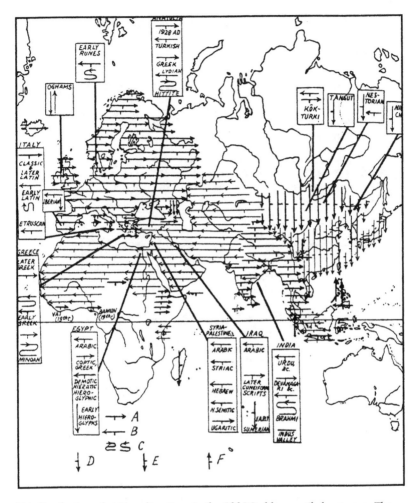

The distribution of writing directions in the Old World up until about 1500. The arrows give the direction in which the characters are written, either in lines or in columns. A short perpendicular line shows where the next line or column comes, so English is an arrow pointing right with a short stroke pointing downwards. A snaking line indicates boustrophedon. The boxed texts give the names of lost writing systems, or indicate developments in areas where the script changed repeatedly. In the latter case the most recent convention is at the top, the oldest at the bottom.

An example of cuneiform script. It means roughly: 'May Ahoramazda preserve this land from the enemy, hunger and treachery.' The vertical strokes present no problems, but the horizontal and diagonal strokes are almost impossible for a left-hander to produce. Cuneiform is the only exclusively right-handed system of writing.

product of Western imperialism and ethnocentrism: imperialism because many ways of writing have been pushed aside over the past five centuries by the relentless spread of Latin script across regions colonized by European powers; and ethnocentrism because the colonizers consistently saw native cultures as inferior and therefore felt their writing habits were unimportant.

Missionaries and evangelists in particular were responsible for propagating the Latin script and eroding local cultures. Today the economic dominance of the Western world ensures that its expansion continues, if rather more slowly. China, Japan and the Arab world do not seem likely to adopt the Latin alphabet, Cyrillic is in glowing good health and the Indonesian scripts and those of Southeast Asia are alive and well, yet Westerners still blithely pass over one inescapable fact: at least half the world's population writes from right to left.

Since it's impossible to point to a consistent preference either way and the vast majority of people everywhere are right-handed, we can safely conclude that there is no such thing as a natural writing direction. It's not significantly harder to write from right to left with the right hand than the other way around, as the Arabs demonstrate. This also means it cannot be particularly difficult to write from left to right using the left hand. There are just a few practical problems that are perfectly easy to solve, with a bit of care and attention – but more of that later.

Nevertheless there is one system of writing that can be produced only with the right hand. It is cuneiform script, whose name derives from the Latin for wedge, *cuneus*, a reference to the shape of the horizontal and vertical strokes that are combined to form the characters. They were drawn in wet clay with a reed stylus that had a triangular profile, so each stroke acquired a small wedge-shaped notch at the start. Cuneiform script was written from left to right, with the notch marking the left end of each horizontal stroke. That is something which is almost impossible to achieve with the left hand.

22

The Weight of the Liver

It starts when we're about two years old and it never stops. We ask 'why?' as soon as we come upon anything that strikes us as unfamiliar or strange. Left-handedness is one example. Virtually since the beginning of recorded time, people have asked themselves why some individuals prefer using the left hand to the right. The more astute among them have immediately realized that there are two questions here. The first and perhaps more intriguing is: why do we have a hand preference at all? Only after answering that question can we ask ourselves why not all of us favour the same hand.

The first person we know to have developed a theory about the cause of one-handedness was Plato. The debate in part seven of his *Laws*, between an Athenian and the Cretan Cleinias, shows that he thought we came into the world ambidextrous but started to favour one hand because of the empty-headed way mothers and carers treat children. This caused 'lameness in one hand'. What precisely it was that had this pernicious influence he doesn't tell us, unfortunately, which worked to the advantage of later adherents of similar theories, since it meant they could all the more easily hitch the world's most ancient scientific authority to their wagons.

Even in Plato's day the grass was greener on the other side of the fence. However superior the Greek philosopher may have felt to the many barbarian peoples of neighbouring lands, he was convinced that the upbringing of children was occasionally ordered better elsewhere – among the Scythians, for example, a much feared warrior people. Scythians, Plato has his Athenian say enviously, are far cleverer than Athenians. They use ingenious training methods to ensure the capacities of both hands and both arms are retained, proving that those who neglect their left side are acting contrary to nature.

To Plato one-handedness was merely a regrettable consequence of a careless upbringing. This cannot be the case, since there were never any truly ambidextrous Scythians and the frequency of right-handedness is roughly the same in all peoples. Aside from the minor impact of taboos, there's hardly any difference between them, whereas we would expect significant variation if hand preference was a product of culture and upbringing.

Since Plato's time only one really serious attempt has been made, by American psychoanalyst Abram Blau, to attribute left- and right-handedness entirely to environmental factors. He carried out his study during and shortly after the Second World War, in the days when behaviourism held the humanities in an iron grip. Before his day, all kinds of mechanistic explanations had come and gone over the centuries.

For as long as the Ancient World lasted, Plato and Aristotle retained their authority. Aristotle had little of interest to say about hand preference. He merely claimed that all movement came 'by nature' from the right and that the left hand was therefore unsuited to operating independently. After the collapse of the Roman Empire people had other things to do than to philosophize about hand preference, so it was the late fifteenth century before someone came up with a genuinely fresh insight.

That someone was Lodovico Richieri, an Italian who lived from 1469 to 1525. He was the first person, although by no means the last, to connect left-handedness with *situs inversus*, a condition in which the layout of the organs in the human body is reversed. Riccchieri, who was active some hundred years before William Harvey described the circulation of the blood, regarded the heart and liver as sources of heat. The heart served the left side of the body and the liver the right. If the liver was for some reason unable to perform its beneficial work to optimum effect, then a person became left-handed. One reason might be that their liver was on the wrong side, on the left.

It was not a particularly strong argument, and fortunately there is no connection between hand preference and *situs inversus*, a disorder with far-reaching consequences. We have been aware of this for centuries, as demonstrated by the work of the curious seventeenth-century Englishman Sir Thomas Browne, a prominent citizen of Norwich. Browne was a typical child of his time, whose studies at Oxford, Montpellier in France, Padua in Italy and Leiden in the Netherlands turned him into a liberally educated scholar in all kinds of fields. He was actually a physician by profession, but one no less at home in the experimental sciences of his time, not to mention alchemy, astrology and sorcery – a true *homo universalis*, who on one occasion even acted as an expert witness for the prosecution in a witch trial, which ended with both ladies convicted as charged.

Although he took seriously a great many things we would now regard as fanciful or based on old wives' tales, his fame was derived mainly from his crusade against superstition. In 1648 he published a book with the wonderful title *Pseudodoxia Epidemica, or enquiries into Very Many Received Tenets and Commonly Presum'd* TRUTHS, *which examined prove but* VULGAR ERRORS. Among the topics he covers in his book are the ideas of his time about left- and right-handedness, and he rightly rejects Ricchieri's *situs inversus* theory on the grounds that the phenomenon is far too rare to account for something as common as left-handedness. *Situs inversus* occurs in only one in 10,000 people and even then it is often incomplete. Sometimes only the organs in the chest cavity are reversed, sometimes only those in the abdomen.

The idea that *situs inversus* had something to do with left-handedness proved astonishingly persistent. As late as 1862 Andrew Buchanan, professor of physiology in Glasgow, presented it as a cause, this time as part of a more general theory of equilibrium.

The liver, he said, is the largest and heaviest organ in our bodies. More of it lies to the right of our middles, so our centre of gravity is slightly to the right. To that extent his argument was correct. Generally speaking, the right half of the body is almost a pound heavier than the left. The theory that he went on to develop was less convincing. Buchanan believed that because of our displaced centre of gravity we suffer from an imbalance for which we compensate by leaning slightly to the left, using our left leg to stand on. This means that the right hand has greater freedom of movement, so people are generally right-handed. Buchanan could explain the existence of left-handed people only by assuming that their centre of gravity was slightly to the right, as a result of *situs inversus*.

You might imagine that such a theory could have been thought up only by someone who'd never seen a left-handed person from close proximity. It's almost unbelievable that a well-trained physiologist of some repute could still think in 1862 that *situs inversus* was common enough to explain left-handedness, and no less strange that he seems not to have taken the trouble to test his theory, eccentric as it was, on the nearest available left-handed person. His ignorance seems all the more woeful given that almost a century earlier, in 1788, incontrovertible evidence had been produced that contradicted Buchanan's assertion. In that year Scottish pathologist Matthew Baillie published in article in which he described a case of *situs inversus totalis*, or a complete reversal of the organs in the torso, devoting an entire passage to the fact that the man in question was right-handed. Despite the fact that the article was reprinted in the journal of the Royal Society of Physicians in 1809, people like Buchanan apparently knew nothing of it. Fortunately Buchanan was soon the target of fierce criticism from all sides – from other physiologists, for example, who had taken the trouble to probe the belly of a left-handed person, or listen to his heartbeat.

From time to time, right up until the end of the nineteenth century, variations on Ricchieri's blood-supply theory kept popping up, often assigning a key role to the arteries under the collarbone, sometimes to other blood vessels. They were all based on an essentially identical assumption: in normal people, blood flows less easily to the left hand than it does to the right, whereas in left-handed people the opposite applies. When it eventually became clear that the causes of hand-preference lay in the brain rather than in the hands and arms, this assumption died a silent death.

One final death spasm was the idea that the blood supply to the left half of the brain was greater than that to the right, enabling the left cerebral hemisphere, which controls the right hand, to become more highly developed. That idea had its blood supply cut off in about 1900, when it became clear that differences in size between the two carotid arteries, each of which supplies blood to one half of the brain, did not consistently favour the left brain. Moreover, those differences were cancelled out altogether by a system of cerebral arteries known as the Circle of Willis.

Mechanistic explanations of one form or another held their ground for four centuries, but in the end none of them came up to the mark.

23

The Morbid Views of Abram Blau

With his 1946 book *The Master Hand*, New York psychoanalyst Abram Blau earned the dubious honour of having launched into the world some of the most stark and insulting ideas about left-handedness ever expounded by a serious and influential scientist.

Blau, a professor and head of child psychiatry at the Mount Sinai Hospital in Manhattan, was a fervent supporter of the then dominant behaviourist movement in psychology, which teaches that practically all human characteristics are determined by environmental influences. Behaviourists regarded concepts such as consciousness, will and character as of little importance, since ultimately even these were shaped by an individual's reactions to impulses reaching him from outside. You could fully understand a human being only by taking a close look at what he did in his environment.

Of course not even the most fanatical behaviourist could deny that inherited characteristics existed. They were known to reside in the chromosomes and in DNA, although when Blau published his book no one yet knew the structure of DNA or how it functioned. Nevertheless, the behaviourists claimed, our genetic material merely confers on us a wide range of possible behaviours. How a person actually develops from birth onwards, what specific behaviour he will display and which characteristics, is determined almost entirely by the formation of habits based on reward and punishment. In Blau's own words: 'Heredity is involved only to the extent that it is a valuable reservoir awaiting to be exploited. The particular choice, direction, and nature of the skill finally achieved are quite outside the province of the germ plasm.'

Reward and punishment are responses to behaviours that people exhibit based on impulses that reach them. Desirable or successful behaviour is thereby reinforced, while undesirable or unproductive conduct is steadily and relentlessly reined in. The world of the behaviourist therefore works according to the principle 'child sees dog, child pats dog, dog bites child, child stays away from dogs in future'. The remarkable thing about behaviourism is not its insistence that rewards and punishments have an influence on our behaviour. Everyone knows that learning by experience is important, if only because it teaches us to keep our hands off hot stoves and strange dogs. What sets behaviourism apart is the way it regards the mechanism of reinforcement and suppression as absolute and downplays the phenomena that accompany it – inheritance, for example, and other biological influences.

As a dyed-in-the-wool behaviourist, Blau naturally attributed the prevalence of right-handedness to social pressure and the formation of habits. In his view it went like this. In the early Bronze Age an enormous amount of time and effort was needed to make tools such as knives and sickles. As a result these were expensive items, cared for meticulously and often handed down from parent to child. Implements like sickles can be used properly only in one hand or the other, and because they were so difficult to replace, no one was going to let a left-hander use them in ways for which they were not intended. So along with tools and the skills required in wielding them, hand preference was passed down from one generation to the next. This effect was further reinforced, Blau believed, by the fact that once the first tools had been invented, one-sided specialization came to have many advantages over ambidextrous development. It would be easier to reach a certain level of dexterity with any given tool if one or other hand were used consistently, and the better trained that hand was, the more easily a person could learn other skills using it.

Blau still had to explain why hand preference was constant throughout history as a cast-iron rule, even among people who had little or nothing in the way of expensive tools. He could not take refuge in Lamarck's belief that acquired characteristics can be inherited, since that view had been discredited by Darwin's *Origin of Species*. Blau claimed the explanation lay in the fact that people pass down their elaborate cultures to a subsequent generation. Unlike other animals, we spend years bringing up our children. Therefore, he wrote, 'culture represents the combination of past social experience and mass learning, [so] each human infant takes over this culture where the previous generation left off. In a literal sense, man's experience becomes immortalized, and part of the property is transmitted to his heirs.' This is plainly true, but it doesn't explain anything. The great value of cultural inheritance, the key to its effectiveness, is its flexibility. Cultures differ in all possible respects from each other and from those of previous generations, whereas right-handedness is always and everywhere the unchanging norm.

Anyone looking at right-handedness the way Blau did is bound to see left-handedness as an objectionable aberration, a pathological consequence of a wilful rejection of the cultural heritage on offer, or the result of a shortcoming of some kind that makes it impossible for an individual to develop into a socially healthy person.

That is precisely what Blau thought. Left-handedness, he believed, was caused by 'an inherited deficiency, faulty upbringing, or emotional negativism'. Of the three, the latter was undoubtedly Blau's favourite. He regarded left-handedness as 'nothing more than an expression of infantile negativism', comparable to 'contrariety in feeding and elimination, retardation in speech, and general perverseness insofar as the infant with meagre outlets can express it'. In adulthood sufferers from left-handedness were no less unpleasant individuals, since their characters were a combination of rebellious stubbornness, secretive superstition, parsimony, an obsession with cleanliness and excessive rigidity in a more general sense. All this – how could it be otherwise – was in most cases the fault of emotional neglect by a loveless 'refrigerator mother' of the kind people at the time held responsible for childhood autism. Blau believed therapy would produce beneficial results in such cases, even though it might be ineffective in eliminating left-handedness once it had been imprinted, leaving it in place as a kind of mental fossil. With small children it was important to stamp out left-handed tendencies with care and tact. Only if the child absolutely refused to switch to its right hand was it better to leave it to stew in its own juice.

Anyone making such bold claims needs to have what it takes to defend them, and Blau did not. His only grounds for rejecting the inheritance theories of his time about the development of left- and right-handedness was the absence of positive evidence. To a certain extent he had a point, since there seems to be no neat Mendelian pattern of inheritance for left- and right-handedness such as exists for fair and dark hair or blue and brown eyes. One thing he overlooked, however, was that his own theory too was built on quicksand, or rather on wild, unverifiable assumptions about life in the Bronze Age.

Moreover, it remains unclear why the development of hand preference, which Blau regarded as culturally determined, would lead to an overwhelming predominance of right-handedness in all times and all places. Perhaps that is the unstated reasoning behind his characterization of its hereditary basis as a 'valuable reservoir': we have a strong innate tendency towards the right. But if this is what Blau means by a reservoir, then he and his behaviouristic thinking fall at the first hurdle and hand preference is simply inherited.

Blau developed his notion of left-handedness as a sign of a disturbed personality based on his experiences with several left-handed patients, but to conclude that their left-handedness was caused by the disorders that brought them to him is absurd. You might as well conclude that wealth is a threat to mental health because poor people so rarely consult expensive psychoanalysts.

In his pitiless conclusions and pronouncements, Blau was guided more by unfounded prejudice than by the products of scientific curiosity. This is clear from the research he carried out among fewer than 400 schoolchildren in an attempt to verify his theories. He found nothing to support his claims, but the failure of his study did not deter him in the least. His book is replete with the kind of heedless, pseudo-scientific bar-room talk that inspired the endless stream of tasteless jokes about psychiatrists and psychoanalysts which continue to do the rounds to this day.

24

Thwacking and Hurling

That heredity must be bound up in one way or another with the origin and continued existence of our one-handedness is undeniable. However, we still have no idea what the nature of its involvement might be, or why right-handedness ultimately became the norm, let alone an explanation as to where all this left-handedness keeps coming from.

We don't even know for certain how long ago right- and left-handedness arose, nor when they settled into the lopsided ratio of nine to one, even though there are a number of indications that the distribution we see today goes back a very long way. Among cave paintings such as those at Altamira, Lascaux and Pech Merle, works of art up to 25,000 years old, we find a large number of handprints. Some are positive, made by first pressing the hand into ochre or some other dyestuff and then against the wall, like a stamp. Most marks of this type are of the right hand. Others are negative, drawn with a piece of charcoal around a hand laid flat against the wall, or by blowing powdered paint across it. These negative prints are mostly of a left hand, suggesting that the right hand drew around the left or served as a platform from which to blow the paint powder at it.

Further clues come from stone tools, which are far older. Many archaeologists claim, based on their shapes and the location of marks of wear and tear, that the people who made these tools an estimated 200,000 years ago included roughly the same proportions of right- and left-handed individuals as we see today. Not all archaeologists agree on this, incidentally.

Evidence going back much further still is provided by the baboon skulls that have been found in the vicinity of the remains of our presumed distant relative *Australopithecus africanus*. We're now talking about some two to three million years ago. Although he was small of

Cave painting of Appaloosa horses with handprints, roughly 20,000 years old, at Pech Merle, France.

stature, *africanus* could successfully hunt baboons because he knew how to use a large bone or sturdy stick as a club. Most of the baboon skulls he discarded have a hole smashed in them with a blunt instrument of some kind, usually on the left side, as if the unfortunate creature was clubbed to death by a right-handed primate. None of this amounts to solid proof, but at the very least there's no evidence that modern man is any different from his ancestors as far as hand preference is concerned.

In the past hundred years or more, two theories have been propounded that are mainly concerned with the question of how it comes about that most of us favour the right hand. Both are based on natural selection. The idea is that during a specific period in the course of the evolution of our species, right-handedness offered advantages over both left-handedness and ambidextrousness in the struggle for survival. Right-handers therefore propagated themselves more successfully in the long run, perpetuating characteristics that occurred in an increasing proportion of the population until ultimately left-handedness and two-handedness almost ceased to exist.

In 1871 that same Thomas Carlyle whom we have to thank for the terms left and right in politics lost the use of his right hand to paralysis. He was 75 years old. Even at that age he had a restless, creative spirit, and the effort involved in learning to eat and especially to write with his still perfectly serviceable left hand set him thinking. Where could that strange specialization of the right hand come from, if the left hand was in theory no less capable? Might it have to do with other asymmetrical properties of the human body?

Of course the obvious thought was that it must have something to do with the heart, traditionally imagined as located in the left side of the chest cavity. There lay the germ of Carlyle's theory. In a fight, a person using his left hand to defend himself and his right to attack is better placed to protect his own vital organs, especially the heart, than a person using the opposite tactic. Right-handedness could therefore have arisen in the mists of time as a result of the invention of the shield, which offers greater protection if carried on the left arm. The right hand would then automatically be called upon to take charge of the complicated business of swiftly striking out while parrying an enemy's thrusts. It goes without saying that those who happened to have more highly developed motor functions in their right arms and hands were at an advantage. Right-handers therefore had a greater chance of survival and so gradually became dominant. Ultimately this led to the well-known ratio of 90 per cent right-handedness, 10 per cent left-handedness.

The time was ripe for such a notion, it seems, since in the same year, quite independently of Carlyle, Englishman P. H. Pye-Smith published exactly the same idea in the medical journal *Guy's Hospital Reports*. This is rather odd, since even in those days a doctor ought to have known that the heart is only very slightly to the left of centre. Any extra protection afforded by carrying a shield on the left side would be infinitesimal; it certainly wouldn't confer a decisive evolutionary advantage. The shield theory, therefore, however attractive in its simplicity, did not stand up to scrutiny. Carlyle's idea could not be correct anyhow, if right-handedness is as old as those baboon skulls suggest. *Australopithecus africanus* may have been quite a dab hand with a length of wood or a large bone, but he definitely didn't invent the shield. Even more fatal to the theory is the fact that later, long after modern man evolved, many peoples never used shields, yet right-handedness is as common among them as it is elsewhere.

Much more recent, and far better thought through, is the argument advanced since around 1980 by American neurobiologist William H. Calvin that he calls his 'Throwing Madonna theory'. The story starts in the far distant past when our ancestors slowly began to develop in the direction of modern man.

Once, millions of years ago, our distant African forebears left their homes in the tropical trees, abandoning the forest for life on the plains. They had probably lived mainly on a diet of fruit up to that point, plus whatever else they could find to eat among the leaves of the trees, including the occasional small animal. In other words their diet was similar to that of the apes today. Whatever their reasons for forsaking their ancestral environment, once down on the ground in open tropical savannah it must have been a good deal harder for them to find sufficient food. Far fewer plants grew there than in the damp and diverse forest. They were able to solve the problem by extending the range of their foraging and by adjusting their diet: more tubers and roots, more small animals. The new skill of running came into play, combined with the well-developed hands of the forest-dweller that had been theirs for millions of years. Those hands were free for the collection and transport of food and soon proved their worth by grabbing animals like mice and rabbits, which lived, like our ancestors, on grassy plains.

This sounds more pleasant than it would have been in reality. An individual with sufficient determination can get the better of a mouse on open ground, but a rabbit is just as nimble, larger and much faster. Like most other animals of the savannah, rabbits are more than a match for humans or hominids equipped with nothing but their bare bands. We are too slow and lumbering, too vulnerable to attacks by large predators. So these early humans cannot have had a particularly meaty diet; they must have been vegetarians most of the time. This in itself is something of a problem, since little energy can be derived from plant material, especially if eaten raw. Being a vegetarian on the ancient savannah meant working long and hard for relatively few calories. Man's prospects did not look good.

Nevertheless, this comparatively slow hunter-gatherer proved himself unprecedentedly energetic. Some two million years ago *Homo habilis* spread from Central Africa across the entire African continent with phenomenal speed and within less than half a million years his successor, *Homo erectus*, inhabited large parts of the globe, fertile or barren, rainy or dry. Eventually he could be found in practically all regions that are warm enough for humans to survive without clothing or heated dwellings. He even settled in areas where it was virtually impossible for him to get by on roots, tubers, nuts and seeds, since they are not productive enough in those environments or simply do not grow in sufficient quantities.

So man must have survived on the flesh of birds and small animals. Meat is a far more compact and digestible form of energy than plant matter, and it's available everywhere. In all environments, whether the frozen wastes near the poles, the tundra or arid sandy deserts with extreme variations in temperature, it's possible to find edible species of animal that are adapted to the prevailing conditions. The largely vegetarian gatherer who from time to time catches a hare must therefore, Calvin claims, have developed in this period into a competent hunter who gathered plant material to supplement a meaty diet.

The surprising thing about all this is that the forerunners of modern man spread so far out across the globe long before hunting implements such as the spear or the bow and arrow were invented. How did our ancestors get hold of enough animals to feed themselves? Calvin suggests we need to look at various differences between our methods of hunting and those of today's apes, which are still limited to their original tropical environments.

One such difference is the impressive arsenal we eventually developed, but here we are looking at a time before weapons were invented. Second, it's often claimed that what made the difference was language, perhaps the most spectacular cultural artefact that humans possess and apes lack, since it's said to have brought about decisive improvements in cooperation between individual hunters. But this does not stand up to scrutiny either. During a hunt, language is not used to any significant extent, as we can see from the success of animals that hunt in packs, such as wolves. They are living proof that ancient techniques like surrounding, driving and wearing to exhaustion can be applied without the slightest hint of language. In fact when people go hunting their conversations usually have nothing to do with the hunt. They talk about the ordinary things in life: relationships, work and money. As soon as they start seriously stalking their prey, they keep their mouths firmly shut. The quarry does not understand what people are saying, but it hears them all the better for that.

There's a third difference between humans and apes, however, and Calvin believes that it's of huge significance. People are capable of throwing accurately. Apes sometimes throw stones, but always more or less indiscriminately. Humans use one arm as a sling of sorts, and this enables them to hit a small target many metres away with considerable force. The capacity to throw in this way gave the otherwise rather unimpressive *erectus* a deadly prehensile claw with a range of at least twenty metres.

The well-aimed throw has a considerable effect on hunting success, and in evolutionary terms it's a skill that can spread through a species extremely quickly. The connection with highly nutritious food is so direct that it exerts enormous selective pressure. Anyone who can throw even a little better than the next individual has a far greater chance of survival and is therefore more likely to pass on his superior throwing ability to a new generation. If, that is, the throwing skill is rooted in something heritable.

This does appear to be the case. Success in well-aimed throwing relies on a powerful capacity for control, which is lodged in the brain. A good throw requires the thrower to develop and implement a plan of action for dozens of muscles, based on his observation of the course and speed of the prey, the weight of the stone in the hand and a great deal more besides. Nature has little to show that can compare to man's accuracy in throwing. Neurons, the electrical switches of which our brains are composed, are not particularly good at such tasks, being relatively slow and imprecise, but if a great many neurons share the same job speed and accuracy increase markedly. Though individually no

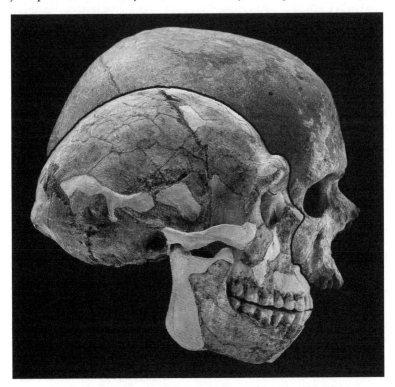

Partially reconstructed skull of *Homo erectus*, with a modern human skull behind it. *Erectus* had a relatively small cranium but was nevertheless a competent hunter.

more than approximations, average results from hundreds of neurons produce extremely accurate estimates, and by having several groups of neurons work simultaneously we avoid the need to spend so much time waiting for intermediate results. The trick is to bring a huge amount of brain capacity to bear. But to do that, of course, you need to have a large brain.

Call it a historical coincidence, or even in the words of Stephen Jay Gould a 'glorious evolutionary accident', but that brain capacity was available. Roughly in parallel with the rapid spread of our early ancestors across the globe, an equally rapid, hugely impressive expansion in their brain mass took place, the first phase of growth that would culminate in the almost one and a half kilos of grey matter housed in the skull of the average modern human. Our new capacity to calculate could not have arrived at a better moment.

Well-aimed throwing is a typically one-sided affair. Its great speed and precision, and therefore the considerable firepower it provides, are achieved by the use of one or other arm as a sling. This means that the required computing power needs to be present in one side of the brain only. Nevertheless, the symmetrical structure of the physical human being meant that the other side of the brain increased in size to almost the same degree. If that capacity was not needed for throwing, it would be available for other skills that conferred evolutionary advantages of their own. Were the throwing capacity to be lodged in both halves of the brain, it would mean extra investment and the use of additional brainpower without much to show for it. After all, a person who can throw equally well with either hand is barely any more effective than someone who can throw accurately only with one or the other. The two-handed thrower may even be worse off, because of the need to choose which hand to use at a point when every millisecond counts.

All this, Calvin argued, sowed the seeds of an increasing dissimilarity in functions between the two cerebral hemispheres, possibly even for the developmental boom in the human brain that eventually produced consciousness, language and all the other characteristics that make us what we are today. At the same time, inevitably, we developed a hand preference.

Of course Calvin's line of reasoning did not emerge out of thin air. He was inspired above all by what we know about the differences in function between the two halves of the brain. It does indeed seem as if, generally speaking, the left cerebral hemisphere is devoted to the kind of tasks in which the precise sequencing of acts and calculations is of importance. This fits in with the fact that humans are predominantly right-handed, since the left half of the brain controls the right half of the body. But now we are starting to get ahead of the argument. Calvin's reasoning offers a fair explanation as to why only one half of the brain specializes in this kind of task, and therefore why we have a hand preference, but his story has nothing to say about which of the two halves would be best. It therefore leaves us with the mystery as to why nine out of ten people have the same hand preference. In other words: what is it that makes the left side of the brain the seat of the throwing mechanism in almost all cases and not the right, or one side in half of us and the other side in the rest?

In his search for an explanation, Calvin joins those who turn to the carrying habits of mothers with children and the position of the heart. As we have seen, most mothers carry their children with the left arm, probably because babies are happiest there and therefore quietest, since a mother's heart is more clearly heard on the left side of her chest cavity than the right. Calvin assumes that *erectus* women were just as active during a hunt as their menfolk and that they carried their offspring with them while hunting, so women with quiet children had a clear advantage. Not only could they move more easily, there was less chance that their positions would be given away by the crying or crowing of their infants as they crept up on their prey. Those quieter children were generally carried on the left, leaving their mothers' right hands free. So it was not the first stone-thrower but the first stone-throwing woman, Calvin's 'Throwing Madonna', who took the decisive step on the road towards the modern human being.

There seems to be little arguing with the first part of Calvin's story. It's both appealing and based on facts. Clearly the forerunners of modern man, once they had climbed down out of the trees, spread across all kinds of terrain without having any special tools at their disposal and started eating more meat. It's undoubtedly the case that brain size increased spectacularly in the same period, and we can throw incomparably better than our closest relatives in the animal kingdom. Finally, it has gradually come to be accepted as a fact that in the vast majority of people living today the left half of the brain is responsible for activities that involve careful planning, considerable speed and correct sequencing. Until someone throws a spanner in the works, therefore, we can assume that his explanation for the existence of hand preference holds water.

The same cannot be said of Calvin's story about why most of us are right-handed. It relies upon assumptions about the social role of *erectus* women that are impossible to prove and against which various counter-arguments can be put. Firstly, the advantage to right-handed women who carry their infants in their left arms ceased to apply once children were no longer taken hunting by their parents. We don't know when this happened. Secondly, we may well wonder whether hunting did not involve such risks to mother and child that women had the best chance of survival if they left it to men and the childless, relying on the solidarity of the group. Occasional examples in the animal kingdom suggest this is the way it worked.

Calvin's idea also fails to chime with the fairly well-substantiated theories put forward by Richard Wrangham, an American biological anthropologist who sees in the sharp contrast between *Homo habilis* and *Homo erectus* the beginnings of cookery. He claims this was an activity that fell to women from the start, partly because they were less mobile as a result of their roles in procreation and childrearing. Based on ideas about precisely the period in which Calvin looks to huntresses as the motor for one-sided brain development, Wrangham places women at the kitchen stove for the rest of eternity.

There is a fourth objection. If the behaviour of a child truly had such a powerful influence on the hunting success of its mother, then the effect on the survival chances of the child itself would be just as significant. In that case, why was the behaviour of the mother rather than that of the child subject to evolutionary pressure? If Calvin is right, you would surely think that quiet children of right-handed mothers had the best survival chances, and therefore tended to survive in greater numbers. It would therefore be perfectly sensible to suppose that modern babies ought to resemble their quietest ancestors and lie absolutely silently in their mothers' arms. To the chagrin of many parents, this is not the case. Babes in arms still shriek, whine and wriggle to their hearts' content.

25

Thinking About Brains

It seems natural to assume that hand preference could have something to do with the difference between the two halves of the brain. Yet it was not until well into the nineteenth century that the connection was made.

Hippocrates, the patriarch of doctors, suspected 2,500 years ago that brains were associated with thinking. That was anything but an obvious assumption. Aristotle, for instance, believed that the warm, beating heart was the centre of everything. He regarded the brain as no more than a system for cooling the blood. Yet even in the ancient world some real, concrete knowledge was available about how we control our limbs, for example. In the third century BC, Alexandrine scholars Herophilus and Erasistratus discovered the nervous system and were able to distinguish between efferent and afferent nerves. The former are the lines of communication that send instructions to the muscles, while the latter respond to external stimuli by transporting signals to the spinal cord and from there to the brain.

Some four centuries later Galen, the Greek personal physician to Roman emperor Marcus Aurelius, performed experiments on animals in which he partially severed their spinal cords in various places. This revealed a great deal about the composition and function of the nervous system. He also knew that the brain had something to do with the mind, but he thought the mind was to be found in the fluid that fills the hollow spaces between the cerebral hemispheres. That idea would persist until the sixteenth century, when the brilliant Flemish physician Andreas Vesalius showed it could not possibly be true. There are those who claim that as early as the first century AD people knew that the left half of the brain controlled the muscles of the right half of the body and vice versa, but we cannot be certain of this.

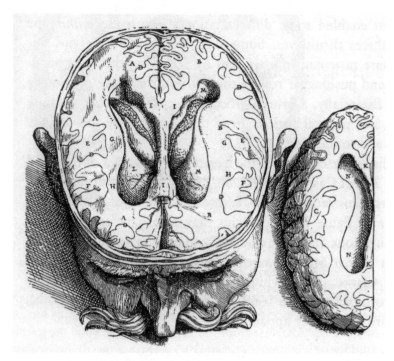

One of the drawings of the brain made by the great anatomist Andreas Vesalius (1514–1564). At the centre are the ventricles, filled with cerebrospinal fluid.

These were impressive achievements, of course, but after Galen hardly anything happened in the field of brain research for almost 1,500 years. The dissection of animals, living or dead, in order to investigate their inner structures had virtually gone out of fashion even in Galen's time and it would be almost unknown for centuries, making way for all kinds of speculative theories that in the end produced hardly anything useful. Still, it's possible this was of little consequence for the growth of knowledge about how the brain works, since across all those centuries a range of dogmas, most of them emanating from the churches, blocked the free development of ideas about the connection between our bodies and our mental processes. The mind, the spirit and the soul were things of a higher order, mysteries that must at all costs remain mysterious. In any case people had few techniques, however primitive, that would have enabled them to study the brain, and there is nothing to be seen in the grey matter itself that could possibly divulge how it works. A damaged brain, like a diseased heart or failing kidneys, caused death. Why this was so inevitably remained a mystery.

The idea that the brain has something to do with thought, emotions and the intangible thing we call the mind nevertheless remained alive down through the centuries. A strange grey lump of matter fills the head, while the face is the location of our most striking features and means of contact with the outside world. Throughout history, people have believed that a person's character and intelligence could be read from his or her face. Aristotle, who felt his own appearance to be no better than ordinary, publicly declared that he had his training in philosophy to thank for the fact that his mind was less ordinary than his face would suggest. It's an extremely deep-rooted notion, as evidenced by the success of comparable ideas presented by Cesare Lombroso in about 1900, to which we shall return.

The turning point came in the eighteenth century, with the success of the Enlightenment. The most important achievement of that school of thought was to dispense with traditional, mainly religious convictions. According to the philosophical innovators of the time, the nature of

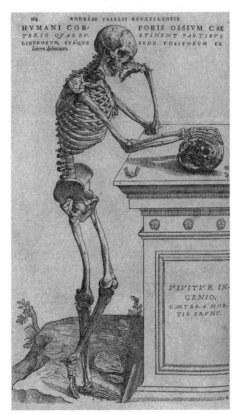

Even Vesalius' skeletons wondered what was going on inside the brain.

things could be discovered through reasoning, aided by empirical science. In other words, people no longer wished to speculate randomly, nor to adhere to an unquestioning faith in the old, respected authorities simply because they were old and respected. From this point on, the message was: look carefully and draw rational conclusions from what you see. Whether or not God and the pope were pleased by this approach no longer really mattered to serious scientists, who were now free to start examining those aspects of humanity that had to do with the mind.

It wasn't long before a great many people started studying the brain, among them the Viennese doctor and anatomist Franz Joseph Gall (1758–1828). He was the first to claim that our brains consist of a layer of grey matter, the cerebral cortex, over a core of a whiter substance. He was also a central figure in phrenology, a movement that had been gaining in influence ever since the mid-eighteenth century and from which many of the terms and concepts used in today's psychology are derived, as well as a whole series of misconceptions and popular fallacies. Phrenology was the first real attempt to link human functions and behaviours with the brain. It was based on the notion that specific human characteristics were anchored in specific parts of the brain, and that you could see from the shape of the skull and its bumps and irregularities how highly developed, or not, certain functions were. Phrenologists were rather casual in the way they went about localizing characteristics. Gall, for example, 'discovered' that the centre for individuality and new ideas lay just above the nose, since, or so he claimed, that particular area was large in the case of Michelangelo but generally small among the Scots.

Phrenologists believed the brain to be a collection of more or less independent organs for characteristics such as aggression, introspection, conscience and inquisitiveness, as well as for language, a sense of time and melody, humour and wit. The larger the organ in relation to others and therefore the bigger the bump in that position on the skull, the more dominant that particular characteristic would be in the person concerned.

Although there were perhaps as many charts of the skull as there were phrenologists, they had one thing in common: all were symmetrical. On the charts drawn up by Gall, Spurzheim and Combe, exactly the same areas are marked out on the left as on the right. Gall seems to have taken it as read that his brain organs were laid down with the same mirrored symmetry as arms, legs, lungs, kidneys, Fallopian tubes and testicles. This is actually rather strange, since even though the cerebrum

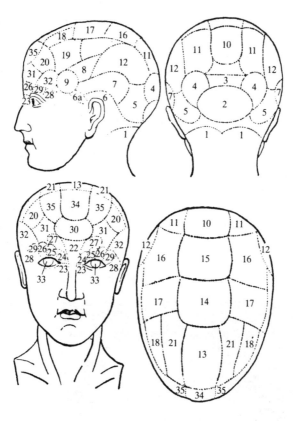

1. Amativeness	12. Cautiousness	24. Size
2. Philoprogenitiveness	13. Benevolence	25. Weight
3. Concentrativeness	14. Veneration	26. Colour
4. Adhesiveness	15. Conscientiousness	27. Locality
5. Combativeness	16. Firmness	28. Number
6. Destructiveness	17. Hope	29. Order
6a. Alimentiveness	18. Wonder	30. Eventuality
7. Secretiveness	19. Ideality	31. Time
8. Acquisitiveness	20. Wit	32. Tune
9. Constructiveness	21. Imitation	33. Language
10. Self-esteem	22. Individuality	34. Comparison
11. Love of approbation	23. Form	35. Causality

Personality areas in the brain according to the phrenologists Gall, Spurzheim and Combe.

consists of two halves that at first sight look symmetrical, the differences between them in the twists and folds on their surfaces are considerable. As a trained anatomist, Gall must have been aware of this – especially since those irregularities were supposedly the features that told us so much.

Phrenology, therefore, was not a true science; it was more a sort of speculative sport for gentlemen of high social standing. Gall simply invented his brain organs and the places where they were located. He didn't discover them – they were not there to be discovered. Yet although the claims of phrenology were extremely shaky and plenty of people were aware of this from the start, the ideas of Gall and others like him were popular until the mid-nineteenth century, and not just among certain groups of neurologists and physicians but among a far broader public. For the educated citizen there could be no more enjoyable party game than determining the characteristics of your household, friends and acquaintances in a 'scientifically sound manner' simply by looking for irregularities on the surface of their skulls. Think of the solace to be

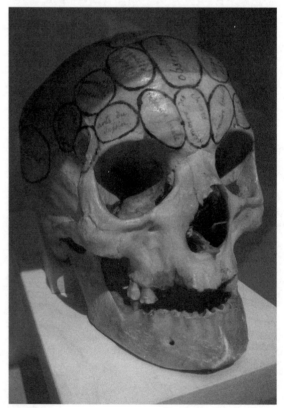

There were as many sets of characteristics as there were heads. On this skull, the 'Protestant saint' of Alsace, Jean-Fréderic Oberlin, indicated the distribution of talents in a way that was all his own. Above the nose, where Gall locates individuality, Oberlin identified a 'memory of things'.

had from the evidence-based notion that you could tell from the broad, flat head of your firm's chief accountant that he was a crafty schemer.

After about 1820 serious opposition emerged to this brisk, unthinking phrenology, encouraged in part by the fact that another researcher, Frenchman Jean-Pierre Flourens, had noted that destruction of different parts of the brains of pigeons often had roughly the same effect. The decisive factor was not so much the location of the damage as the amount of brain mass affected. Flourens argued therefore that characteristics and functions cannot be located at specific places. The brain works as an indivisible whole. Although he rather blithely passed over the fact that birds' brains are not at all the same as human brains, his work marked the beginning of the end for phrenology. As is often the case with scientific plausibility, the pendulum swung to the other extreme. From this point on the localization of brain functions was regarded as nonsense. Anyone who continued to believe that certain parts of the brain had specific tasks to fulfil could expect to meet with a great deal of scepticism in scientific circles.

This is probably the reason why French family doctor Marc Dax made little impression at a congress in Montpellier in 1836 with his discovery that patients with brain damage on the left side often had trouble speaking and understanding speech. This was the basis of his conjecture that linguistic functions are located primarily in the left cerebral hemisphere. He was unable to find a publisher for his work. Within thirty years the situation had changed again. In Paris in 1861, Paul Broca (1824–1880) announced the discovery of the area of the brain that is named after him. Broca reported that time and again speech problems arose when precisely that area was damaged, whereas linguistic comprehension, memory and other functions related to language might remain intact. The area he had identified must therefore be specifically connected to speech.* Within ten years of the publication of Broca's work, Gustav Fritsch and Eduard Hitzig discovered the bands across the middle of each hemisphere that were responsible for the control of the limbs and other movable body parts, brain areas now known collectively as the motor cortex.

This changed everything. The localization of brain functions turned out to be possible after all, as the phrenologists had always said it was, if on the wrong grounds. While it wasn't the case that entire personal

* In 1865 Dax's son succeeded in having his father's work published, but by then the credit had gone to Broca. Not entirely unfairly, since Broca had a far more precise idea than Dax as to where speech functions were located.

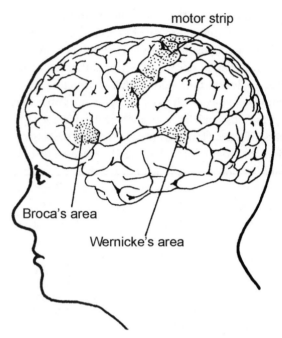

motor strip

Broca's area

Wernicke's area

The left half of the brain, showing the areas identified by Broca and Wernicke that are involved, respectively, in speech production and language comprehension. At the top is the motor cortex, used to steer moveable body parts. Immediately behind it lies the somatosensory cortex, which processes signals sent from the skin to the brain by the sense of touch.

characteristics or qualities such as creativity, mendacity or perseverance had a fixed location, far simpler, often rather abstract functions certainly did. Relatively primitive things such as registering the place where the light falls on the retina, sensing where you are being touched, bending and stretching an arm or finger, or moving the jaw muscles to create a chewing motion can today be traced to activity in precisely defined areas of the cerebral cortex. Complex character traits, by contrast, are the ultimate result, at a very high level, of brain processes working together, combined with our upbringing, the laying down of experiences in our memories and various external stimuli. There's no way for us to recognize in the final result the basic components that go to make up aspects of our characters, any more than we can recognize in the graceful motion of a ballet dancer performing *Swan Lake* the nerve impulses that race through her body, stretching or contracting the muscles at precisely the right moment.

The functional symmetry that phrenologists took for granted turned out not to exist, or to exist only in part. Broca's area is located in the left brain only. The equivalent area in the right brain has nothing to do with speech, and the same goes for a good number of other areas that have since been discovered. This meant that the two halves of the brain could no longer be regarded as equal, which inevitably raised the question of which half was in charge.

One myth was thereby exchanged for another, since almost immediately the idea took hold that the left half of the brain was dominant. The mere fact that most people are right-handed was enough to suggest that the mechanisms controlling that hand, known to originate in the left brain, were more highly developed. This made it all too tempting to regard that cerebral hemisphere as superior in a general sense.

Furthermore, all the newly discovered specialist areas that were found in only one half of the brain were in the left half, and they had to do with typically human and therefore highly esteemed capabilities such as linguistic and – later – arithmetical skills. The first and most important discoveries were Broca's area, which is mainly involved with the production of speech, and an area described by German Carl Wernicke in 1874 that seemed to be engaged mainly in language comprehension. Although the motor cortex, responsible for controlling physical movement, is pleasingly symmetrical, occurring in both halves of the brain, German scientist H. Liepmann showed in the early years of the twentieth century that the left half nonetheless performs a unique function in coordinating complex actions.

Everything conspired to make the left cerebral hemisphere seem central to the human mind, richly endowed with high-level, typically human functions. It was exclusively responsible for the body's more demanding physical acts and charged with the most important intellectual feats: the production and comprehension of language, arithmetical competence and, via the right hand, writing. For a long time the right half of the brain seemed to be there to make up the numbers. It put people in mind of an arid, sparsely populated desert, where nothing happened that was worthy of note, nothing at least that helped us with anything beyond the demands of bare survival.

For a while there was fairly general agreement that the left cerebral hemisphere was the overall champion, but as to the function of the other half, at least two different beliefs emerged. One described the right half of the brain as largely a fallow reservoir of brain capacity that offered possibilities as yet unknown. It could take over the functions of the other half if necessary, so it might be capable of almost anything. From an evolutionary point of view if no other, this is a rather problematic story. It would be quite extraordinary if such an energy-demanding, astonishingly complex organ had existed for millions of years without being used. Nevertheless, this belief prevailed for long enough to introduce the misconception, as optimistic as it is ineradicable, that we use only 2, 10 or 20 per cent of our brainpower.

The other belief was that the right hemisphere duplicated the left, serving as a kind of backup system. It was not empty but full of slumbering copies of the apparatus active in the left half, ready to jump into the breach on demand. It certainly is the case that our bodies have duplicate systems that provide us with spare capacity. We have two kidneys, two lungs and two Fallopian tubes, for example, mirror images of each other, even though we could get by perfectly well with one. Yet none of these duplicate systems behaves the way people imagined the right half of the brain to behave. Both work at 100 per cent capacity from the start; we don't have a kidney or a lung that sits idly waiting to be called upon should the other cease to function. This alone makes it extremely improbable that we have brains largely made up of an idle reserve mechanism.

Not until the 1950s did it begin to become clear what was going on in the right half of the brain, initially as a result of work by researchers including Henry Hecaen in France and Britons Oliver Zangwill and Brenda Milner. A series of right-brain tasks came to light, including the interpretation of visual information and later the identification of people by their faces, musical abilities such as the recognition of melodies, and spatial orientation.

These new insights were brought to us by means of new research techniques. Previously almost all available knowledge had been drawn from what were known as natural experiments, a rather high-sounding way of describing random events. A person might be brain damaged in a fall, by a blow to the head, or after a stroke or cerebral infarction that incapacitated him or her in all sorts of different ways. Once the patient had died, sometimes many years later, researchers could open up the skull and see which parts of the brain had been destroyed. By studying the location of the damage and the symptoms that accompanied it, and comparing the effects in different patients, a picture gradually emerged of which bits of the grey matter were essential for which tasks.

It was better than nothing, but it did not do an enormous amount to help research along. Accidents, strokes and infarctions disable random and often quite large parts of the brain that don't neatly coincide with a presumed arrangement of brain regions devoted to specific functions. They often produce a complex jumble of disabilities that is difficult to interpret, and they never strike in exactly the same way twice, which makes it extremely hard to compare one patient with the next.

Many people are too badly affected to take part in studies, and on top of all this comes the eternal problem that dogs this kind of research: the fact that two things coincide does not mean there's a direct causal connection between them. You cannot simply take the fact that with a specific form of brain damage a specific function is lost and conclude that the location of the damage indicates the place where that function resides. After all, if you take a wheel off a car it can no longer move, but that doesn't mean the car travels on just that one particular wheel.

Until the middle of the twentieth century there were two other possible ways of learning about the structure of the brain. One was the surgical removal of part of the cerebral cortex. Extraordinary circumstances aside, this route was blocked in the case of human beings. Scientists therefore had to resort to animal experimentation, with all its limitations and uncertainties. The only alternative method consisted of the direct electrical stimulation of parts of the cortex during brain surgery. This technique did in fact lead to the discovery of a number of areas that have to do with voluntary movement and sensory perception, and it is still of great service to surgeons in the process of determining exactly where they need to make incisions, but until recently it was too crude to tell us anything about less concrete and primitive functions. A technique called deep brain stimulation is now yielding intriguing results, but the procedure involves a great deal of risk, so it can be applied only in the most serious cases of brain damage.

Eventually new techniques were discovered that made it possible to allow the two halves of the brain to work independently. They included the astonishing split brain operations carried out by American neuropsychologist Roger Sperry and a number of colleagues in the mid-1960s. The *corpus callosum* was severed, as were any other, smaller connections, so that the two cerebral hemispheres could not come into contact with each other except by a tortuous route through the brain stem.

This may sound like the work of a mad scientist, but Sperry was no Dr Frankenstein. He had been presented with patients affected by a severe form of epilepsy. They suffered terrible seizures caused by powerful, spontaneous waves of electrical impulses that bounced back and forth between the two halves of their brains via the *corpus callosum*, rather like water sloshing in a bowl that's being shaken. The idea was that by breaking the connection the waves of impulses would be prevented from building up such devastating strength. Animal experiments showed no other detectable consequences, so Sperry began separating the two halves of his patients' brains.

It worked. Sperry's operations brought considerable relief to many hopeless epileptics, but at the same time they offered a unique opportunity to address each cerebral hemisphere separately, by supplying information to one half of the retina only. Research of this kind confirmed beyond doubt that essential tasks involved in the organization and production of language generally take place in the left half of the brain, while the recognition of images mainly happens in the right half. If one of his patients saw a house with the right side of his retina, and therefore with the right half of his brain, he often had no difficulty pointing with his left hand to things that are typically associated with houses yet was unable to say what he could see. Finding the word that refers to a house is a task of the left brain, and that half had no idea what the patient was seeing. Conversely, if the word 'house' was presented to the left side of the retina – and therefore to the left half of his brain – he was able to read it but in most cases could not point to the appropriate picture. In order to recognize an image of the thing referred to by the word he was reading, he would have had to engage the right half of his brain, which was impossible in any direct way because the two halves were no longer connected.

A less drastic though still deeply invasive treatment was what's known as the Wada test, after its inventor the Japanese-Canadian brain surgeon Juhn Atsushi Wada. In the years immediately following the Second World War, Wada worked in Japan with brain-damaged ex-servicemen who were being treated with electric shocks. The therapy brought them some relief, but it also caused further damage to the men's memories and linguistic abilities. After much trial and error, Wada hit upon the idea of using amobarbital (sodium amytal) to shut down the left cerebral hemisphere, where those functions were located, before the shocks were administered. Then fate decided to give Wada a helping hand. He was sent a young man who had been shot in the head and as a result had descended into an endless sequence of epileptic attacks. One fit immediately sparked the next. This no longer resembled uncontrolled sloshing back and forth in a bowl, it was more like a pan of water brought to a furious boil. The patient did not regain consciousness even for a moment. It seemed nothing could be done, until Wada tried breaking through the chain reaction by anaesthetizing the left half of the young man's brain. The convulsions ceased.

Wada had not only chanced upon a way of reducing the damage inflicted by electroconvulsive treatments and gaining a grip on seemingly uncontrollable epilepsy, he had unintentionally invented the Wada test. Anaesthetizing one half of the brain gave him an unprecedented opportunity to take someone with an undamaged brain as his test subject and communicate with just one or other of the cerebral hemispheres. He was able to confirm and expand upon the findings of Sperry with his split-brain patients. If the left half of the brain was incapacitated, the test subject was temporarily unable to count or talk and had great difficulty understanding spoken or written instructions. If the right half was sent to sleep, the person in question proved unable, for example, to sing.

On the other side of the world, a couple of years before Wada set to work, a German neurologist had made an important discovery. Klaus Conrad nursed and performed tests on some 800 men with bullet wounds to their brains in a German field hospital in the final years of the war. It may sound harsh, but gunshot wounds are popular among brain researchers, since they cause neater, more clearly defined brain damage than accidents, strokes or infarctions. He discovered something extraordinary. Since Broca's time almost a century earlier, scientists had believed that the language centres were located on the same side of the brain as was responsible for the control of the preferred hand, which in the overwhelming majority of right-handed people meant the left side. It was assumed that the brains of left-handers were reversed, so that both these functions were located in the right brain, an assumption that grew out of the old idea that the two sides of the brain were each other's mirror image in some way. Echoes can be detected here of ancient ideas about left-handedness and *situs inversus*.

Although Broca's Law, as it was known, was generally accepted, it had never been tested. Left-handers are relatively unusual in any case, and people with aphasia and not too much other brain damage are rare. It's therefore very unusual for a neurologist to come across a person who is both left-handed and aphasic but otherwise healthy. It would take a world war to bring enough of them together for reliable research. Conrad found himself in the middle of that war, and he discovered that a considerable percentage of the left-handed sufferers from aphasia he was treating had been shot in the left side of the head, just like the right-handed aphasics brought to his hospital. Clearly the widespread assumption about reversed brains in left-handers was false.

This produced a new insight that would persist into the twenty-first century. The brains of left-handers were not the mirror image of those of right-handers; instead they were rather less markedly lateralized. More than two out of three left-handed people have their language centres in the left half of their brains, just like right-handed people. The remainder fall into two groups, one with 'everything' on the right side and another whose linguistic capacities are divided between left and right. People in this latter group have a slightly better chance of avoiding aphasia after suffering a serious head wound, although the advantage is only slight. One might hope for their sakes that they never have the opportunity to discover by such a route just how unusual their brains are.

People would not be people if the discovery of functions in which the right brain seemed to specialize had not immediately become the basis for new myths and misconceptions. The apparent contrast between right-brain tasks and the specialist functions of the left brain – counting, arithmetic and the production and processing of language – were quickly transformed by all kinds of experts and semi-experts (in ways that percolated through even to women's magazines) into a concrete antithesis between a cool, calculating, analytical 'person' who resides in the left half of our heads and a warm, emotional, holistic character ensconced in the right half. From there it was only a short step to the idea that a person's character was ultimately determined by whether the right or left half of the brain held the reins. Artistic, musical and highly emotional people were thought to be right-dominant, whereas the more analytically inclined cold fish clearly had brains in which the left side was boss.

This fitted perfectly with the archetypal social distinction between arts and sciences, so perfectly in fact that practically everyone failed to notice the strange consequences of such a train of thought. The boffin or nerd is a clever but quiet, shy, socially awkward figure, whereas the creative, hard to pin down, humanities-oriented libertine thrives on social intercourse. The latter holds forth, polemicizes and writes romantic letters, dramatic poems or passionate literature. Supposedly left-oriented scientists seem far less able to handle that archetypal left-brain function, language, while right-dominant types rely upon it. For decades no one pointed out the contradiction here, or suggested it might pull the rug out from under the theory of dominance. Instead the differences between the two halves of the brain were simplified into a table of opposites, an ostensible order that was permitted to be inconsistent and hardly ever tested against reality. It seems we just can't help ourselves.

No less inevitable was the link made between the dominance model and hand preference. Left-handers were stamped as generally less verbal

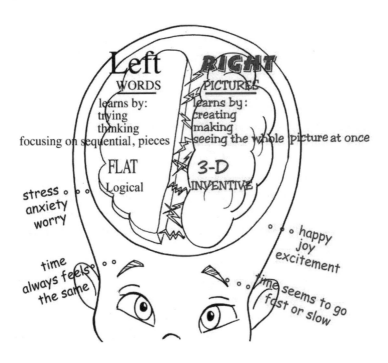

Promotional material for a course that promises a more harmonious life. Telling left from right has proven too difficult for the organizers of the course.

but more creative than right-handers and above all more geared towards the visual. For many left-handed people, traditionally put down as clumsy, this may have been a comforting thought, but that's all it amounted to in the end. No evidence has ever been found for their greater artistic endowments. Meanwhile the development of new techniques, which offer far better opportunities for glimpsing the workings of the brain, has made it seem a good deal less likely that such evidence will ever be found. Those techniques did not arrive until the final decades of the twentieth century.

As Jean-Luc, still a little confused after his madcap ride in the nose cone of the rocket, crawled into the space station he looked up, straight into the moonface of the commander.

'Mr Picard, welcome on board the International Space Station. How was Baikonr? Is everything still such a mess there in Kazakhstan? They'll never learn, you know; those Russkies need a tsar to order them around. Glad you could come. We always have a good laugh with the French.'

'Thank you, er, I'm happy to hear that,' Picard answered, somewhat taken aback. 'I hope I can make myself useful here.' And he paddled along behind the man who would be his boss for the next two weeks as they made their way to the residential area of the space station.

'Absolutely. Don't you worry. But first I'd like to show you something.' The commander pointed towards a small porthole in the side wall of the module. With a sweep of his arm he nudged the weightless, unsuspecting Picard in the right direction.

'Ow! Merde – sorry, I still have to learn to slow down in time.' Crossly, Picard rubbed his cheek, which had just made rather abrupt contact with the porthole. But then he forgot the pain, mesmerized by the view. Outside, surrounded by millions of sparkling, pin-prick stars, was a huge dark ball with a bright golden aura: the earth, with the sun concealed behind it. It was night down there.

The commander tapped the window and pointed: 'Look. See that patch of light? No, further to the right: Paris! I thought you'd like that.'

Picard looked. A patch of light was all he could see. But in his mind's eye an image unfurled of cars, buses and people in a great swarming mass. He could imagine the noise, the sight of the brightly lit shops and the theatre doors just opening, or perhaps already closing, the dizzying complexity of all the things that go to make up a city. What was left of all that from here, less than four hundred kilometres above the earth's surface? A fuzzy splodge.

The view that a novice space traveller would have of Paris from an earth orbit bears rough comparison with our image of a healthy brain functioning normally. Inspired mainly by rapidly advancing digital technology, one new instrument after another appeared on the scene in the late twentieth century, enabling us to catch a glimpse of what goes on inside the skulls of people with normal brain function, without any need for surgical breaking and entering, even without anaesthetics. The most important new technique dates from the early 1990s and is known as fMRI, or functional Magnetic Resonance Imaging. It was initially called Nuclear Magnetic Resonance Imaging, but that name fell into disuse in the medical field because it created undesirable associations with radioactivity.

Where brains are hard at work they use a great deal of energy in the form of oxygen, supplied to them in the haemoglobin that fills red blood corpuscles. In the more active areas of the brain, blood flow quickly becomes more intense than in areas where little is happening, and the hungry neurons extract so much oxygen from the corpuscles streaming towards them that their haemoglobin becomes oxygen-poor, giving it different magnetic characteristics from oxygen-rich haemoglobin. This difference can be measured, creating a profile of brain activity.

The measurement process is at first sight rather like making an x-ray photograph, but without the use of dangerous rays. When oxygen-rich haemoglobin comes into contact with a magnetic field, it forms a very weak opposing magnetic field. One might say it works a tiny bit against the external magnetic field, whereas oxygen-poor haemoglobin forms a barely detectable magnetic field of its own in the same direction, thereby slightly increasing the strength of the external field. So if a body is exposed to a uniform magnetic field on one side, we can measure, on the other side, point by point, how much of the strength of that field remains. The 'shadows' shown up by this process indicate that somewhere between the source of the field and the point of measurement lies tissue the magnetic forces are finding harder to penetrate. The deeper the shadow the more resistant the brain matter.

The contraption in which all this takes place is the MRI scanner, a huge, powerful, ring-shaped magnet into which it's possible to slide an entire human body. Once the subject is inside the machine, a base measurement is made, or a map of the brain at rest. This is achieved by producing at great speed, with the help of sophisticated computing techniques, a large number of adjacent magnetism profiles, in other words magnetic photographs of cross-sections of the head. A sequence of slices, so to speak, is made from left to right and another from front to back. The two sets of images are then combined to create a three-dimensional picture of the brain.

If that brain is now given something to do – such as recognizing a word or an image, counting to ten or solving arithmetic – the areas of the brain involved in that task are activated immediately. The scanner again records a complete three-dimensional series of slices, this time of a thinking brain. The differences between that set of images and the base measurements indicate the places in the brain that have been activated. Their intensity shows the degree of activity.

So we first measure the structure of a person's brain, then the activity taking place in it. We know exactly what tasks the brain is performing, since they have been given to the brain by the people making the measurements. This is what the 'f' in fMRI stands for: the measurement and localization of an activity or function.

There have been huge advances in fMRI since the final years of the twentieth century. Every month, newspapers and magazines feature new images of the brain with active areas lit up in fetching colours, usually with captions saying that we now know where human brains recognize words, experience joy, generate a sense of embarrassment or whatever else it might be. This is a great deal more than we used to be able to see, and it is indeed truly impressive, but at the same time, all we can expect from this technique is a rough, coarse-grained map. As yet we have taken only a small step on the long road towards unravelling how the brain works. Even with the best scanning techniques in the world we see only the kind of thing our space traveller saw on the surface of the earth: indistinct splodges. They tell us which areas are involved in certain tasks, but not exactly what they do or why. We see Paris, but not the streets swarming with people, cars and buses that make Paris what it is. We cannot see the cafes, shops and theatres, let alone understand where all those people and cars are going, or what happens in those cafes and shops.

Initially fMRI results brought nothing new to light that could be of relevance to hand preference, which is hardly surprisingly, since people were by then a little tired of studying left- and right-handedness and there was a lack of exciting new ideas. Left-handers weren't even regarded as valuable test subjects. Their brains were known to differ a fair bit from the right-handed standard, not just when it came to manual dexterity but in the positioning of the much investigated linguistic functions. Their non-standard patterns of brain activity would only make it unnecessarily difficult to come up with an unambiguous interpretation of data from groups of test subjects.

This changed in 2009, and again it was the Max Planck Institute in Nijmegen, along with the Donders Institute for Brain, Cognition and Behaviour in that same Dutch city, that produced new research on the subject. Studies showed that various visual functions are linked to hand preference. In recognizing faces, left-handers use more neurons in the left brain than in the right. This shattered one cornerstone of what we thought we knew about the geography of the brain, since facial recognition had for decades been a textbook example of a capacity located exclusively in the right half of the brain.

It had already been known for some time that the brains of left-handers are rather less lateralized than that of the average right-hander. This explains why left-handed people have a slightly lower risk of becoming aphasic if the left half of the brain is damaged. What the researchers in Nijmegen discovered was that the brains of left-handers differ from the right-handed norm in many other ways as well. In recognizing faces, bodies and chairs, right-handed people deploy mainly the right cerebral hemisphere, whereas left-handed people make more use of the left. In most cases it's correct to talk of facial recognition as a right-lateralized function, but this cannot be said of left-handers. Both at certain locations and more generally, the two halves of their brains differ rather less.

This raises all kinds of interesting questions. For example, are the memories of left-handed people organized differently and what does this tell us about the way memory works? One theory of memory is that the remembered meaning of a word – a concept – is not simply an abstract thing that exists independently of all else; rather it becomes rooted in the physical characteristics and sensations of the bearer of the concept, if at all possible. This is known as embodied cognition. In concrete terms it implies that the meaning of a verb of action such as 'lick' or 'kick' consists in essence of the activation of the area of the brain that comes into play when we are in reality about to lick or kick something, shortly before the motor cortex sends the appropriate command to the muscles. In other words, the comprehension and understanding of the concept 'to kick' consists of carrying out, up to a certain point, the planning phase of the act so named.

If the theory of embodied cognition holds water, then it would be logical to expect left- and right-handed people to differ in their ways of understanding and remembering those verbs that refer to actions carried out by the hand, such as 'pinch', 'throw', 'pick up' or 'draw', aided as they are by the pre-motor areas for their preferred hand. This turns out to be the case. Results of fMRI tests show that with right-handers the

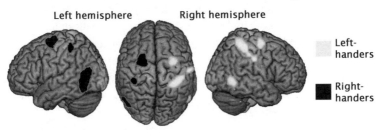

Left hemisphere   **Right hemisphere**

Left-handers

Right-handers

The areas of the brain that light up in left-handers and right-handers when verbs for actions involving hand movements are recognized. The meanings of such words seem to be stored not only in the areas of the brain that are concerned with motor functions but in the side of the brain that controls the preferred hand.

relevant areas in the left cerebral hemisphere are highlighted and with left-handers the corresponding areas in the right.

Add to this the remarkable phenomenon discovered in 2009 by Daniel Casasanto – that people respond more positively to things in the real world that appear on the same side as their preferred hand than they do to things on the other side – and little remains of the image we once had of the differences in how the brains of left- and right-handed people are laid out. Those differences can sometimes be far greater than we used to think and they seem to relate to a far broader range of functions, perhaps even to all functions. Strong lateralization, with tasks that are entirely reserved for one half of the brain, seems to be far less in evidence than we once imagined. Lateralization was once seen as a monument not only to our knowledge about our brains but to the difference between us and other species. That monument is now being shaken to its foundations.

Despite the apparently larger, more varied and more polymorphic differences that have been discovered between the brains of left-handers and those of right-handers, left-handed people remain as ordinary and inconspicuous as ever. The most important conclusion may therefore be that our brains are far more flexible than we once thought.

All this is not to say that the beliefs of the second half of the twentieth century were entirely incorrect. On the contrary. There's no reason to doubt that in the majority of cases the right cerebral hemisphere is concerned with such things as spatial orientation, the creation and recognition of melodies and the interpretation of images. The left is usually better at counting, arithmetic, keeping track of time and processing language.

Yet we need to take into account that many of the functions we recognize as such are not single entities at all but the result of a meshing together of a wide range of smaller tasks and capacities. To think, to understand, to read and write – these are brief, simple terms for unfathomably complex processes. It would be quite strange if all the processes involved in such higher functions were located in either one cerebral hemisphere or the other.

The functions we can positively identify as located in a specific place in one or other half of the brain will tend to be relatively straightforward, abstract processes, dozens of which go to make up what we think of as higher functions. Perhaps the location of a given function is not even particularly important. The overall difference in approach between the two halves of the brain may be far more relevant. It may be that the difference between our cerebral hemispheres has less to do with the specific processes that reside in them than with how those processes happen. In other words, they may differ more in their way of working than in what they do.

To find one possible indication of this we need to turn our attention back to our hands. We can all make a fist, lay one hand flat on a table or drum a tabletop with our fingers. People with brain damage can do all these things too, so long as they aren't hampered by severe paralysis. All of us, including brain-damaged people, can easily mimic those three acts in a given order after they have been demonstrated to us once: ball fist, lay hand flat, drum fingertips. Unless, that is, there is damage to the left side of the brain. The performance of each act still presents no difficulties, but getting them in the right order does. The revelatory thing here is that whether the patient happens to be left- or right-handed, this difficulty is manifested not only by his or her right hand, which is controlled by the damaged left brain, but to an equal degree by the left hand, controlled by the undamaged right brain.

Clearly the problem does not lie in hand-control. It seems those damaged parts of the left brain are involved not so much in the performance of the acts themselves as in making them happen in the required order, with either hand. Healthy people and people with damage to the right brain generally have no problem at all with simple sequencing tasks.

This suggests that the planning and organization of complicated undertakings, in other words the compiling of a programme of activities that must be carried out in a specific order, is a speciality of the left brain. It's precisely this capability that lies at the root of William Calvin's ideas about the causes of hand preference. Ever since Liepmann's discovery in the early 1900s that the left cerebral hemisphere is involved in complex motion in both halves of the body, this breakthrough has been only a short step away. It also tallies with the likelihood that brain damage on the left side will cause linguistic problems, since the processing of language is all about putting together and plucking apart sentences and words, structures that have to be compiled in precisely the right order. In fact language processing involves a highly complex arrangement of levels and subsidiary tasks, each of which has its own strict rules about order and sequence that fall under the heading of grammar. Each phrase, whether spoken or heard and understood, requires the processing of a vast pile of sequence-sensitive data with such astonishing rapidity that it doesn't delay us at all.

We can perhaps safely conclude that the left and right cerebral hemispheres both do their work in their own way, sometimes alone, sometimes in collaboration. The left brain is primarily skilled at aspects of tasks that have to do with counting, language production and comprehension, timing and consciousness of time – all tasks in which sequencing is important. The right brain seems to be more active in the recognition of situations and in associating them with each other. A competitor in a quiz who unfailingly recognizes any song from its first chord can probably attribute that talent primarily to the right side of the brain, which identifies the unique combination of timbre, pitch and so forth that occurs only in the opening bars of that particular number. If the competitor then taps out the rhythm, the left brain becomes fully involved, since he or she is engaging in an activity organized in time.

These differences in the way the two halves of the brain work seem to be little more than tendencies. Strong tendencies, certainly, but they don't amount to an absolute distinction. There's room for considerable variation, for a greater or lesser contrast, for more or less obvious lateralization. It seems left-handedness can somehow be a consequence of this variation, but it needn't be. We shouldn't forget that as far as the separation of functions goes, the great majority of left-handed people show the same pattern as almost all right-handed people. Nor should we forget that despite the effectiveness of modern technology, we still have little firm knowledge about the brain.

The two cerebral hemispheres differ not only in what they do and in how they function but anatomically, which is to say in their material form. A great many parts of the brain are larger on one side than the other in most people. Sometimes they differ in shape as well. In left-handers these differences seem on average slightly less pronounced, and in a few cases the proportions are actually the reverse of those found in most right-handers. When it comes to brain anatomy, left-handers also demonstrate rather more variation, but again this is no more than a tendency and the majority are no different from the average right-hander in the physical shape and composition of their brains. There may therefore be a weak connection between hand preference and brain shape, but there is not, to use phrenological terminology, a bump in the skull representing hand preference.

There is another possible anatomical cause, which strictly speaking lies outside the brain. It has sometimes been suggested that hand preference arises from differences in the nerve bundles that connect the brain with the arms and legs. The left half of the brain directly controls not only the right arm and hand but the whole of the right side of the body, and vice versa. In practically all respects, that is. The connections between one half of the brain and the muscles or, for example, the sense of touch on the opposite half of the body take the form of a thick bundle of nerves that runs from the brain down the spinal column and from there to all the extremities on that side. But there are also connections between each cerebral hemisphere and the body on the same side. This bundle of nerves is nowhere near as thick, although it does mean that if one side of the brain is damaged, the side of the body it controls won't necessarily be completely paralysed and totally numbed.

From each half of the brain, then, two bundles of nerves lead to the spinal column, a thick bundle that connects with the opposite side of the body and a thin one straight down. The thick bundles cross each other just below the brain, reaching the side they control immediately above the spinal cord. These two thick bundles, despite the fact that each has roughly the same amount of body to control, can differ in size considerably. The same goes for the thinner bundles. In eight out of ten people

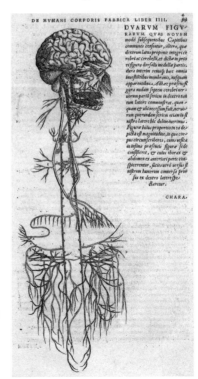

Vesalius, *Fabrica* (1543). The main features of the nervous system are shown here, including the loose bundles of nerves that lead to the arms and hands and the long connections that lead via the spine to the torso and legs.

both the thick and the thin bundles that run to the right side of the body are thicker than their counterparts that serve the left side. Eight out of ten; a very suggestive proportion. You might almost suspect that hand preference arises out of a better infrastructure that makes traffic between the brain and that half of the body smoother and more precise. Almost, because this turns out not to be the case. There is no connection at all between hand preference and the difference in the thickness of these nerve bundles.

26

Animal Crackers

When in the course of the nineteenth century it gradually became evident that in humans real differences exist between the two halves of the brain, a great many neurologists immediately hurled themselves upon the animals of the woods and fields in order to find out whether perhaps there was evidence of similar specializations. The crude neurological techniques of the time made it impossible to use healthy humans in research, so scientists were limited to examining people who happened to have suffered brain damage, or carefully tinkering with those who for one reason or another needed to undergo brain surgery. There were fewer difficulties with animals, which were enthusiastically experimented and operated upon, sometimes with astonishing results. Contrary to expectations, animals often turned out to have a preference for one paw over the other and to stick to that preference with great determination.

In one experiment rats were put into a cage with a tube attached, in which lay a tasty treat. The tube was mounted in an extreme corner of the transparent front of the cage, up against the side, so that the rat had to use the paw on that side to reach into it. Some rats stubbornly kept trying to use the wrong paw. They clearly had a marked preference, so much so that it didn't even occur to them that they could easily get at the treat with the other paw. There were less dogmatic rats too, for whom it made no difference where the tube was placed. They had no hesitation in using whichever paw was convenient.

Even more remarkable results were achieved when rats with a strong paw preference had the part of the brain controlling that paw deliberately damaged. After a few days they were able to move their preferred paw again, but it functioned less well than before. Nonetheless, the rats continued to favour it. This seemed very similar to the way people respond. Whatever may cause hand or paw preference, once it's become established there's an extreme reluctance to switch.

Yet however surprising the results of experiments like these may be, they produce little of relevance to human hand preference. It turns out to be almost impossible, for example, to find a specialization in one half of the brain of an animal, whether ape or frog or anything in between, that resembles the kind of specialization we see in humans. In birds such as sparrows and canaries, the left brain does seem to be more involved in singing than the right, but bird brains differ so fundamentally from those of mammals – and therefore people – that this doesn't tell us a great deal. Another problem is that hardly any criteria are imaginable that could be used to measure a bird's preferences. In rats, reaching for treats serves as a relevant task, but how do you measure foot or wing preference in a sparrow or a seagull?

To a greater degree even than experiments involving people, animal experiments are hard to interpret. Appearances can be deceptive, increasingly so the more an animal differs from us. The renowned scientific journal *Nature* reported in 1996 that 'hand' preference had been found in an amphibian for the first time: the European common toad. Researchers had stuck pieces of wet paper over the animals' noses and mouths, or pulled small balloons over their heads, and watched to see which of their front feet ('hands') the animals used in their initial attempts to get rid of the annoying obstruction.

The Italian-Australian research team reported sensational results from its toad-baiting. The creatures seemed more like people than any other species. No fewer than six out of ten were right-handed, while only one in six favoured the left. The remaining quarter had no clear preference. This was extraordinary, since if chance determined hand preference then a quarter would be left-handed, a quarter right-handed and the rest indifferent. Toads were demonstrating a distinctly uneven distribution, just like humans, such that the left-handed, at around 15 per cent, were proportionately almost as rare as left-handed people. For a short time it seemed as if lateralization, the presumed source of our hand preference, could be traced back to the common ancestor of humans and toads.

Unfortunately the scientists had spoken too soon. Tomio Naitoh of Shimani University in Japan and Richard Wassersug of Dalhousie University in Halifax, Canada, took up their pens to disabuse the readers of *Nature*. Toads, they wrote, have a habit of removing poisonous and indigestible food remains via their mouths. They do this by vomiting up their entire stomachs and wiping the stomach lining clean with a front hand. Since a toad's stomach is asymmetrical, with a shorter membrane on one side, it always hangs out of the right side of its mouth, which makes it easier by far to reach with the right hand. It was very likely that the toads were reacting to foreign bodies on their faces as if they were cleaning their stomachs. You could still call this a hand preference, but its cause lies in the structure of a toad's intestines, not in its brain. So much for the right-handed common ancestor of man and toad.

A few years earlier, in 1990, other researchers had claimed that rhesus macaques were right-handed, based on the fact that the bones in their right arms were usually slightly longer and stronger than those in the left. But oddly, the researchers added, the difference in males became smaller with age and the difference in females larger. They cautiously concluded that something else might be at work here. Two years later New Zealander Rachel Baskerville identified what it was. Changes to the skeleton that coincide with age and sex often have a hormonal cause. In humans, for example, asymmetries were found long ago in the vicinity of the shoulder, a result of the fact that testosterone acts slightly more on the bone on one side than the other. Testosterone levels fall in men over the course of a lifetime, whereas in women they increase slightly, a pattern neatly reflected in monkey bones.

The differences between limbs in rhesus macaques could also be a side effect of small irregularities in the symmetry of the torso as a whole. In people with back problems caused by malformations or by a tilted pelvis, one leg is often found to be slightly shorter than the other, by way of compensation. Whatever the precise reason for the difference found in the monkeys' arms, it probably has nothing to do with paw preference.

A multiplicity of animal experiments has nevertheless produced some results. First of all it's clear that paw preference does exist in mammals, but in a way consistently different from the picture in humans. In animals with a paw preference, the right-pawed group is always roughly similar in size to the left-pawed group, while around half of individuals lack a clear preference. We never see anything like the uneven distribution so characteristic of man.

An even more intriguing result is that breeding for paw preference, in contrast to breeding for colour and all kinds of other characteristics, has proven impossible. It was tried for many years with mice to no avail. In generation after generation the picture remained the same: there were always roughly as many left- as right-pawed individuals plus a large group that showed no preference. This proved, if nothing else, that paw preference is not passed down simply according to Mendel's laws of inheritance. If we persist in thinking that paw preference in animals has something in common with hand preference in people, then the laboratory rat has only succeeded in making hand preference slightly more of a puzzle than it already was.

27
Other Asymmetries and Preferences

Although in most cases we couldn't say straight off whether the people around us, even those we know well, are left- or right-handed, there is one category of individuals concerning whom careful records are kept: sportspeople. Left-handedness is a far from trivial advantage to those engaged in sports in which contestants enter into single combat, such as tennis, fencing and boxing, and indeed baseball, which in essence comes down to a duel between pitcher and batter. The reason is not hard to discover; it lies in the simple fact that left-handers constitute a small minority.

Take training. Most of the time spent by anyone seriously involved in a sport is devoted to rigorous practice. Since most people are right-handed, each player has a right-handed opponent to train with around nine times out of ten. For left-handers this is somewhat awkward initially, since they're faced with asymmetry, but after a while they become entirely used to playing against a right-hander. If they find themselves facing a fellow left-hander in a match, they have little difficulty coping; the situation may be slightly unfamiliar, but they're suddenly in their element. At last the constellation is symmetrical. For well-trained left-handers it therefore matters little which hand an opponent prefers.

Not so right-handers. Like left-handers they train with right-handed people nine times out of ten and usually this suits them fine, but when they occasionally have to take on a left-hander they're doubly disadvantaged. They're forced to engage in an asymmetrical battle for which they're poorly prepared, against an opponent who's a dab hand at dealing with this type of asymmetry. No wonder that in the official rankings of boxing, tennis and fencing champions, left-handers are significantly over-represented, as they are on the lists of top scorers in baseball.

The sports world tells us something else too: hand preference is not the only kind of asymmetry people encounter. A minority of soccer players are left-footed. They prefer to use the right foot to stand on while kicking the ball with the left. Such players are in demand because they can shoot from angles that the majority of footballers find hard to deal with, and because they're in a minority they're just that bit more unpredictable and difficult for right-footed players to mark.

As well as right- and left-footed soccer players, there are some who can hold their ground equally well with either foot. They are fairly few in number but more common than two-handers. If people exist who can truly do everything just as easily with either hand – while still being reasonably dextrous – then they're as scarce as hens' teeth. 'Two-footedness' is encountered relatively frequently, possibly because the tasks of a preferred foot are a good bit simpler than the performances we expect from a hand. In the end it's really only a matter of ball control, the equivalent of accurate throwing. Feet don't have any other skills, if only because all our toes do is help us walk, or keep our balance when standing still, tasks that involve both feet equally. A single specific skill, however complex, can be mastered by the non-preferred side of the body as long as we train long and hard. Left-handers who have been forced to learn to write with their right hands are living proof of this, as are left-handed musicians who play right-handedly.

Yet most soccer players stick anxiously to their preferred foot, and many never manage to train the other to anything like the same standard, so a kick with the wrong boot often fails completely. This was demonstrated on one occasion by the legendary, pig-headed and exclusively left-footed Dutch footballer Willem van Hanegem, renowned for his curved balls, as they're known. Once, in his prime, when he was selected to take a penalty that could have made him the country's top scorer, he took it with his right foot out of sheer perversity – and missed by a mile. Vintage Van Hanegem.

If we have both a hand preference and a foot preference, might there not be other, similar kinds of asymmetry elsewhere in the human body? Aside from our arms and legs there is just one other body part that serves mainly to carry out active tasks so delicate or complex that it allows us to speak of a preference asymmetry: the tongue. True, we have only one, but it is symmetrical, and each half is controlled by the opposite half of the brain. It's called upon to perform some of the most difficult jobs we face.

First of all the tongue is responsible for an efficient preliminary treatment of the food we put into our mouths. It has to ensure that nothing escapes the grinding force of the teeth while itself remaining out of their reach. When this occasionally goes wrong we're instantly reminded why it's important. Its second task is to form speech sounds, which rely upon the subtle curves and differences in shape assumed by the tongue one after the other at impressive speed, in close coordination with movements of the lips, lower jaw and vocal cords. Considering that most of us can speak at a rate of 180 words a minute without too much difficulty, and that on average a word consists of four or five different sounds, it's clear that some fairly impressive acrobatic feats go on between our teeth.

Some people do indeed seem to have a preferred side. In normal circumstances we're not aware of having a tongue preference, but it's easy to identify. The trick is to place one side of the tongue, then the other, gently between the molars and hold it there while singing your national anthem. The tongue side that is free when the words of the song take the least effort to sing and the result sounds best is the preferred side. Sadly no information is available as to how many people have a clear preference, what proportion of people are right- or left-tongued, or whether there is any connection between that figure and the proportion of left- and right-handers.

All the other body parts in which some kind of preference can be seen are used not so much to interact with our surroundings as to detect and observe them. In other words, they perform sensory rather than motor functions. The most prominent among them are the eyes and ears, and those are also the only two about which reasonably reliable facts on the matter are known. In both cases people have a clear preference.

We generally use one of our eyes for seeing and the other for measuring; one looks while the other mainly serves to measure the angle by which what we see differs between the two eyes, so that we can judge distance. This is known as depth perception. In many people, one eye works better than the other. Those of us who wear glasses or contact lenses know this, since the lenses are usually of different strengths. You might expect people to prefer to use their better eye as their seeing eye, but it turns out this is not so. It seems eye preference is settled early in life and remains the same even if the accuracy of one or both eyes changes, as often happens around the beginning of puberty. There seems to be an echo here of the stubbornness with which laboratory rats stick to their paw preference through thick and thin.

Eye preference can be tested in all kinds of ways, but most methods are less than reliable because measurements are distorted by hand preference. There's not much point in taking account of which eye someone uses to aim a gun or a bow and arrow, for instance, since hand preference determines how such weapons are held and therefore which eye is used. What does work reasonably well is to observe how a person looks through the lens of a telescope or microscope, or the viewfinder of a camera. In the case of small children a pleasingly simple test has been developed. All you need is a hollow tube, a rolled up newspaper for instance. You look through one end of the tube and ask the child, who is moving around the room freely, to look back at you through it. The eye that appears at the other end of the tube is the child's preferred eye. Although various studies into eye preference have produced markedly different results, it does seem that almost everyone consistently favours one eye or the other. Roughly two-thirds of us give priority to the right eye, the rest to the left.

Ears are another matter entirely. We don't know whether we have a true ear preference for the simple reason that it's almost impossible to think of a test that could determine which one it is. There's certainly no test that could do so without any interference from other preferences, such as right- or left-handedness. All the same, it's clearly a point worth examining, since in detecting sounds, two perfectly healthy ears turn out to work differently. Whether one or the other is dominant depends on what sort of sound they're picking up. In general it seems we hear sounds that have to do with language, in other words speech, better with the right ear than with the left.

The most important means by which this has been studied involves what are known as dichotic hearing tests. Simply put, the test subject wears headphones and different words are sent to each ear at the same time. The subject then has to repeat what he or she has heard. It often turns out that words fed into the right ear have stuck in the memory rather better than those delivered to the left. If the sounds are unconnected with language, musical tones for example, then the results are the opposite, though the difference is not very great.

We need to treat these results with some caution. Dichotic tests place quite exacting demands on the test subject. It's therefore perfectly possible that the effect arises in part for spurious reasons. In an attempt to produce an acceptable answer, a subject may for example come up with clever strategies that ultimately have nothing to do with the nature of the material presented. The differences do seem significant enough to be taken seriously, however, so we need to ask what the causes could be. We are not dealing merely with differences in sensitivity, since a less good ear hears both speech and other sounds less well.

The only remaining possibility is that this has something to do with differences between the two halves of the brain as they process sound. Each ear is directly connected to both cerebral hemispheres. We might therefore assume that the shorter connection has precedence, either because it's quicker or because it's less subject to disturbance along the way. Yet this is not the case. Speech sounds are received better by the right ear even though the centres that process speech are usually on the left. Conversely, the right half of the brain is more involved in processing melodies and yet we hear musical sounds better with the left ear. It therefore seems as if the nerve pathways coming from the opposite ear drown out those that run to the closer side of the brain, rather in the way that adjacent telephone lines can disturb each other. Perhaps we don't so much have an ear preference as one ear that's better at hearing certain sounds, depending on the place in the brain where that kind of sound is processed and the architecture of the specific nerve circuits that connect our ears with those processing centres.

If ears distinguish between linguistic and non-linguistic sounds, might not something similar happen with vision, in the sense that letters and words are more effectively viewed using one half of the retina and non-linguistic, or perhaps we should say non-symbolic pictures and patterns are better processed by the other? The suggestion seems natural, but it's difficult to prove or disprove. Attempts have been made, but so many interfering factors and uncertainties come into play that we wouldn't be justified in drawing any firm conclusions.

The existence of such a broad range of asymmetries naturally raises the question of whether there's any connection between them. Is a left-handed person also left-eyed, left-footed and left-tongued? And if there is such a connection, does it have to do with the way the brain is organized? Could it be, for example, that in many respects left-oriented people have a distribution of functions between the two halves of the brain that approaches the opposite of that found in the rest of us? One person who was deeply convinced that a strong correlation existed between the various forms of left and right orientation was an American called Beaufort Sims Parson. In 1924 he published a foolproof method of determining innate left-handedness with a machine he had built himself, the manuscope.

It was developed to detect a person's hand preference objectively and reliably. After all, not only are people less than entirely trustworthy in answering questions about their hand preference, they are sometimes pushed in the direction of right-handedness by their environment, even more so in Parson's day than in our own. This happens mainly, although not exclusively, at school. At home too, since most parents are right-handed, children are shown how to do things with the right hand and many appliances are designed to be used by right-handers. Conversely, it's possible that a child of left-handed parents who is essentially right-handed might behave in some respects like a left-hander.

This kind of social pressure is entirely absent when it comes to eye preference. We are generally not even aware that we have such a thing. Parson not only believed it was connected to hand preference, he saw it as the cause. This idea had already been enjoying a degree of popularity for some years. He therefore felt that with his manuscope he had created an Egg of Columbus. Once eye preference had been determined, Parson reasoned, it would be clear whether the person in question was left- or right-handed.

Unfortunately for Parson and the other adherents of the eye-preference theory, it later transpired that their assumption was simply wrong. A majority of us are right-eyed, just as a majority are right-handed, but the proportions are quite different. Moreover it turns out there are not significantly more left-eyed people among the left-handed than among the right-handed. People have a hand preference and an eye preference, but it seems the two preferences develop entirely independently of each other.

More or less the same applies to feet and legs. The right-legged seem to be in the majority, but again the proportions are different from those we see with hand preference, nor is there any direct, demonstrable connection between leg preference and hand preference. It may even be that leg preference is a product of chance, like paw preference in animals. Some people say our preferred leg is the one we set off walking with, but what does such a thing really demonstrate? The only activity that brings to light a difference that looks even remotely like hand preference is soccer, and most people can't kick a football particularly accurately with either foot. It may well be that the roughly 30 per cent of us who are left-footed are matched by a roughly equal number of right-footed people, with a large group in between who have no preference at all. A proportion of 30 per cent is not too far from the one in four that characterizes a chance distribution.

As far as the tongue is concerned, our knowledge is extremely limited. It would be a little bizarre to ask large numbers of people to sing songs or recite poems with half their tongue between their teeth, and who can say whether or not it would render up anything useful?

28
Tallying Up

What is it exactly that's so irresistibly funny about the eternally squabbling cartoon duo Tom and Jerry? Surely it has a great deal to do with the inexhaustible supply of unpleasant surprises, like the door with a brick wall behind it. Less funny was the discovery that medical and psychological breakthroughs of the second half of the nineteenth century were rather like the opening of exactly that kind of door. It was clear that hand preference had something to do with specialization by the two sides of the brain, but it would be a long time before the brain became accessible for direct investigation.

So people resorted to indirect methods, using psychological and statistical research to chart the traits that accompany left-handedness more often than we would expect based on chance alone. Scientists hoped this would at least help to determine whether hand preference was inherited and whether left-handedness was an abnormality, perhaps one that pointed to other problems. Although this would not lead to an explanation of the origins of hand preference, it might perhaps teach us something about the reasons why to this day one in ten newborn babies are left-handed. Over the years this approach has produced an impressive pile of reports and research data.

For practical reasons many such studies focus on groups that already differ from the norm in some way: children at schools for the disabled, people undergoing hospital treatment, residents of care homes or other institutions, prisoners, and pupils with all kinds of problems who have come to the attention of school doctors and psychologists. One major advantage of this approach is that you don't need to go looking for subjects. You can concentrate instead on groups that tend to live in organized contexts, in buildings that are easily accessed by the staffs of universities, institutes and hospitals. Their lives are regulated by systems in which it's normal to be investigated, so studying hand preference is a relatively effortless business, and they generally have little to do, which makes them ready and willing to take part in innocent-looking research projects. It's a welcome diversion, something to distract them from their less than cheering daily routine. Furthermore, all kinds of medical and psychological data are available on such people, so it's easy to make comparisons. All this has the incidental but far from trivial advantage that the research can be kept relatively cheap.

There's another reason for concentrating on these particular groups, one that's at least as important: the existence throughout history of preconceived ideas and prejudices about left-handed people, beliefs reinforced by the discovery of lateral specialization in the brain. Left-handedness is a minority phenomenon, and because aberrations are negative it demands to be corrected almost by definition. This is one of the many ways in which our tendency to think in opposites, in 'we' who conform to the norm and 'they' who are different, influences our attitude to our fellow human beings.

The crudest example of supposedly scientific research in the recent past that was in fact based primarily on unshakable prejudice – and not only in relation to left-handedness – is the work of Lombroso, who believed he could reliably read off character traits and aspects of a person's make-up such as intelligence, trustworthiness and criminality from external features, including the shape of the face and head, body posture and, naturally, hand preference. Around 1900 he turned his attention to left-handedness and promptly found significantly elevated rates of it among criminals, especially female criminals. His categorization of left-handers as among the more civilized kinds of miscreants was presumably meant as a compliment. No fewer than one in three swindlers and racketeers were left-handed, he said, whereas in respectable people the rate was one in twenty. Murderers and violent assailants, the great skull-measurer claimed, display a far less striking tendency towards left-handedness, with figures of no more than about 9 per cent.

Lombroso's conclusions turned out to be incorrect whichever way you look at them and his work has long since been consigned to the graveyard of science, but other consciously or unconsciously biased research persisted. In 1911, for example, psychiatrist E. Stier, on behalf of the German high command, carried out research into left-handedness in the army. Stier had published work on the subject before and he believed it was an inherited characteristic particularly common in primitive peoples. This not only reveals the man's prejudice, it shows he was the kind of academic who

The faces of German criminals illustrated in Lombroso's standard work *L'Homme Criminel*. They include shoplifters (A), murderers (E), pickpockets (H) and burglars (I). The rest are all crooks of various other kinds. According to Lombroso there must be around twenty left-handers here.

paid little heed to facts and figures. We should therefore not be surprised that on average he found only 4 per cent of people to be left-handed, with the exception of the most stupid soldiers, among whom he discovered more than three times as many. Abram Blau, American psychiatrist and scourge of the left-handed, warned young parents as recently as 1961 in a newspaper article: 'Don't let your child be a leftie!' Even the great British child psychologist Cyril Burt had an attitude that could hardly be described as free of value judgements. In his monumental 1937 work *The Backward Child* he characterizes left-handers as follows: 'They squint, they stammer, they shuffle and shamble, they flounder like seals out of water. Awkward in the house, and clumsy in their games, they are fumblers and bunglers in everything they do.' Talk about prejudice. A measured assessment is the last thing you'd expect after that.

Lombroso, Stier and Blau aren't exactly shining examples to the rest of science – no sensible person would take their tall tales seriously today – but Burt is a different matter, since although controversy once surrounded the reliability of his twin studies, he was generally regarded as a meticulous researcher and is still held in high regard. That a serious scientist could allow himself to be carried away to such an extent by traditional prejudices, without noticing that in daily life he came upon so few of those floundering bunglers and squinting stutterers, demonstrates how perfidious prejudices can be and how hard it is to break free of them. Stereotypes undoubtedly had an influence both on the choice of research topics and on the interpretation of the data generated.

Over the past 70 to 80 years many hundreds if not thousands of studies, large and small, have examined the rates of occurrence of left-handedness in specific groups of people. The results are by and large disturbing, at first glance at least. Whatever the abnormality present in the group studied, there almost always seem to be more left-handers than would be expected based on chance alone: stutterers, dyslexics, children with special educational needs, sufferers from hay fever, asthma, allergies and other autoimmune diseases, epileptics, breast cancer patients – in every case they included a remarkably high proportion. There's even a small study that suggests a clear link with alcoholism, another in which an unmistakable correlation is found with criminal behaviour, and a paper in which left-handedness in the more deprived neighbourhoods of a major American city is associated with heavy smoking.

Just as one swallow doesn't make a summer, one study doesn't make a hard and fast rule, even if the media often imply that it does. The results of this kind of research do something else entirely. They indicate the likelihood that an observed correlation between two phenomena is the result of pure chance. If that likelihood is small enough, then the scientist has a significant result and will write an article for an appropriate scientific journal, so that fellow scientists can work out whether or not what he or she says is true. This certainly does not amount to firm proof that the correlation exists in reality, let alone that one phenomenon causes the other. The publication of research results can better be compared with a decision by the police that there is sufficient evidence to arrest someone as a suspect in a criminal investigation. It indicates a reasonable suspicion of involvement, but no more than that.

No study is any more reliable than the assumptions on which it's based. Errors of logic that creep in while an experiment is being set up, or while the data are being processed and interpreted, make the results as worthless as a car with wooden cylinders under its gleaming bonnet, or a computer with woollen circuits. Worse still, as long as no one brings those mistakes to light, the results will mislead. This is the origin of countless spinach-causes-cancer stories. Carelessness in carrying out experiments has a similar effect, as does the use of instruments that don't do exactly what a researcher thinks they do, or reliance on assumptions that don't square with reality. In short, it's not at all easy to design and implement even the simplest study, and even the most careful, meticulous stickler of a researcher can run into problems with the most dangerous assumption of all: that he's testing a representative sample.

In practice it's hardly ever feasible to involve every member of a relevant population in research. Only more or less long-term, hugely expensive population surveys can do that, and they are the exception. They're usually set up to look at common, fatal diseases such as breast or cervical cancer, and even so they are extremely rare. Researchers usually work with a manageable number of people who are presumed to be representative of the population as a whole, but it's difficult to be sure whether this is truly the case, if only because no one knows exactly what a normal person is.

The only convincing method is to compile a random sample of the entire population, one large enough to win the approval of statisticians. In practice this is hardly ever possible. For reasons of money, labour and time, the sample group is usually so small that it's barely statistically valid, and selection is anything but random. All too often scientists resort to students and easily accessible populations in homes and institutions, leaving the general run of the population wholly out of account. An additional problem is that people have to give informed consent before they can be allowed to take part. This too colours the picture.

The key to achieving a dependable result despite all these limitations is repeatability. This is one of the most important reasons why researchers are obliged to publish accurate and detailed accounts of how they developed and carried out their work. Such publishing requirements mean that faults in the design of studies can be pinpointed in retrospect. This is exactly what happened to the scientists who thought they'd discovered that toads are right-handed but had failed to realize that for anatomical reasons the test they'd invented was incapable of proving any such thing. They were rapped on the knuckles a few months after their work was published.

If other researchers using different research subjects with the same approach and the same methods of implementation come up with extremely similar results time and again, chances are those findings are actually meaningful. The likelihood that researchers have made exactly the same unnoticed mistake independently of each other becomes smaller every time the test is repeated. If they make a different mistake on each occasion, then in the long run the distortions this creates will tend to cancel each other out. A stable average result will emerge in which we can have a reasonable degree of confidence.

Bearing all this in mind, if we look at the studies that have linked left-handedness to all kinds of disorders and ailments, it immediately becomes obvious that there are countless reasons for scepticism. One major problem, for example, is the lack of any unambiguous and generally accepted definition of left- and right-handedness. Some researchers divide their subjects into two groups, others work by categorizing them as left-handed, right-handed and mixed-handed. Sometimes a person is regarded as left-handed if he performs, or says that he performs, just one task with his left hand, sometimes only if he does literally everything left-handed. Researchers usually present lists of questions, but sometimes people actually have to carry out tasks, while in other cases various dubious dexterity tests are used to compile hand preference scores. One study may involve a checklist of only three tasks, whereas another may introduce ten or twelve criteria. A researcher may refuse to take account of which hand a person uses when writing because of the possible influence of social pressure, while a fellow researcher takes the writing hand to be the most important indication. In theory at least, a person who is defined in one study as right-handed might quite possibly be categorized in another as left-handed.

There are a great many flashes in the pan in the form of studies that are never repeated. This is understandable. Research on hand preference is not a high priority, and many test subjects need to be recruited before you have enough left-handers to render up reliable figures about small differences in the relative proportions of left-handed and right-handed people.

The effort required to recruit all those left-handed research subjects is a real snake in the grass. You need a group large enough to allow you to extrapolate from your findings. As a result the issue is very often turned on its head. Instead of investigating whether a given disorder occurs remarkably often or remarkably rarely among left-handers, a readily available group with a particular characteristic – prisoners, dyslexics, cancer patients – is studied to see whether it contains more left-handed people than are found in the general population. It's by this route that practically all the reported connections with ailments, aberrations and inadequacies have been found. On the few occasions when tests have been done in the opposite direction, it's never proven possible to find any significant correlation. Time and again, the results produce the paradoxical conclusion that people who suffer from a particular disorder are more likely than average to be left-handed, even though left-handers are not any more likely than the general population to suffer from that same disorder.

Over time a number of facts have emerged that give us something solid to grasp in our search for the causes of hand preference (and indeed for its ultimate consequences, but more of that later).

One such firm fact is that although the results of research are never exactly the same when tests are repeated, many studies large and small show that left-handedness is slightly more common in men than in women. Ten as opposed to 9 per cent; 14 as opposed to 12 per cent: those are the kinds of figures produced.

We can also be fairly certain that people with brain damage, or people in whom brain damage is suspected, no matter how slight, are rather more likely than average to be left-handed. Based on this finding, scientists have tried to discover whether there is a connection between left-handedness and problematic births to older mothers, and these tests quite often show slightly elevated numbers. All in all, we cannot entirely dismiss the notion that left-handedness may sometimes be the result of damage to the brain.

Brain damage certainly cannot explain all cases of left-handedness, however. For that there are simply too many right-handers with evidence of brain damage and far too many left-handers in whom no abnormalities can be found. We must therefore assume that left-handedness can also arise in some other way.

Another fact that emerges repeatedly is that the percentage of left-handers in the population falls sharply with age. Variations on the saying 'anyone who isn't on the left at twenty has no heart; anyone who's still on the left at forty has no brain' have been attributed to a range of famous men, from Winston Churchill and Georges Clemenceau to Aristide Briand, but always with reference to politics. In some bizarre way it seems to apply to hand preference as well, since between the ages of twenty and forty the proportion of the population that calls itself left-handed declines significantly.

There is also firm evidence for an inherited component, one indication being the demonstrable connection between hand preference in parents and children. Couples in which one partner is left-handed are roughly twice as likely to have a left-handed child. (With two right-handed parents the chances are about 10 per cent.) If both parents are left-handed, the figure doubles again, approaching 40 per cent. We might initially think this could be a product of imitation, of setting an example that is then followed, but that's extremely unlikely. Firstly because the majority of left-handers have two right-handed parents and the majority of children of left-handed parents are right-handed, and secondly because there are indications that in at least some cases hand preference develops before birth. Around 5 per cent of unborn babies suck their left thumbs in the womb, and this seems later to translate into left-handedness. Parents couldn't possibly have any influence on this kind of behaviour in an unborn child.

Another indication that inherited factors are at work is that left-handedness occurs roughly twice as often in monozygotic twins as it does in the products of single births. A thoroughly mystifying fact is that the same goes for dizygotic twins.

It's therefore clear that we need to focus our search for the causes of hand preference on hereditary influences and minor mishaps. These are the basis for the two most important theories doing the rounds today, the genetic theory put forward by Marian Annett and the hormonal testosterone theory propounded by Norman Geschwind.

29

Genetic Left-handedness

In the twentieth century several attempts were made to explain left-handedness by reference to genetic differences. Most came to nothing. This was mainly because the distribution of left- and right-handedness did not coincide with the laws of inheritance developed by Gregor Mendel, a shy Austrian monk who by studying the peas in his garden unravelled the fundamentals of how hereditary characteristics are passed on from generation to generation and the patterns according to which they manifest themselves.

Mendel made his discoveries based on the external appearance of pea plants. He knew almost nothing about what was going on inside them. We now know that every ordinary cell of a plant, animal or human contains inheritable material, almost all of it in the cell nucleus. It's organized into chromosomes, tiny coils of immensely long spaghetti-like strings of protein and DNA, pieces of which, known as genes, contain a recipe for making one or other protein, or for controlling the activities of other pieces of DNA. Together the genes of all the chromosomes in an organism form what has been described as a book containing the instructions for making it. A living organism is by far the most complicated thing in the world, so it's a monumental work, sometimes hundreds of volumes long.

In the case of human beings there are 23 volumes. In other words, the nucleus of each of the cells in our bodies contains 23 different chromosomes. Since they all occur in pairs, the nucleus of every normal human cell contains a total of 46, and in every case except one, each pair consists of two copies of the same chromosome.

The 23rd pair is the exception, since only in women are the two chromosomes the same. That's what makes a woman a woman; we are talking here about the sex chromosomes x and y. Women have two

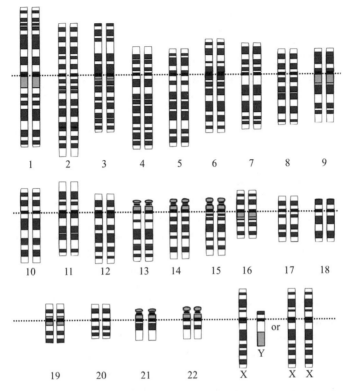

Stylized representation of the chromosomes in the nucleus of a human cell. The ordinary chromosomes, known as autosomes, are numbered from the largest to the smallest. The sex chromosomes are called X (female) and Y (male). At the bottom right are the two possible pairs. A man has the pair xy, a woman the pair xx.

copies of the x chromosome, whereas in men one copy of the x chromosome is paired with one copy of the y chromosome. The possession of a y chromosome is what it means to be male, from a biological perspective. It's responsible for the inheritable characteristics that make a man different from a woman.

Were it the case that at conception two normal cells, one from the mother and one from the father, simply merged, then the resulting child would have not two but four copies of each chromosome. The next generation would have eight, its children sixteen and so on until in a relatively short time the cell walls would literally burst under the pressure of billions of chromosomes. Instead, reproduction is exclusively the task of special reproductive cells: a woman's eggs and a man's sperm. At the point when they are ripe and ready to do their work, they contain only one copy of each chromosome. When an egg cell merges with a sperm cell, the resulting cell has two copies of each chromosome, sometimes including an XY pair, sometimes an XX pair.

A child therefore inherits one complete set of chromosomes from its father and one from its mother. Still, it's plain to see that children are never simply a half-and-half mix of their parents. There are two main reasons for this.

The first is the existence of gene variants. Many genes occur in different forms, with different effects. Sometimes the results are slight variations, such as differences in hair colour, or in the colour of the skin and eyes. Sometimes certain variants cause illness; they are the source of genetic disease. The variants of a gene, also known as alleles, are subject to rules of precedence. Some, described as recessive, will be pushed aside if the chromosome with which they are paired contains a different variant, taking effect only if there are two of them. Other alleles are dominant, in other words: what they say goes. Since a father and mother each contribute one copy from each of their pairs of genes to their child's chromosomes, combinations of characteristics will occur in the child that are not seen in either of its parents. So although children resemble their parents, they are not born with a straightforward combination or a wholly representative sample of their parents' characteristics.

The second cause of variation between children with the same father and mother is a phenomenon known as recombination. Evolution is a knock-out race, governed by chance, and it has rendered up a system of reproduction that guarantees as much variation as possible. This gives a species the best chance of survival, since where a multitude of types is created, there are always one or two that can flourish in the circumstances that happen to prevail at any given time. Recombination means that during the process that will eventually produce reproductive cells with one copy of each chromosome, the two copies are first thoroughly whisked together. The single chromosome that is passed on is therefore not exactly the same as either of the copies in the cells of the parent but rather a random mix of the two. This ensures that children don't exactly resemble their parents in every respect, and it also means they are a bit like just about all their forebears. The shuffling of cards that we call recombination means that from each of a person's many ancestors, one or two characteristics are likely to show up many generations later. Recombination is also responsible for the fact that aside from gender, no one precisely resembles a brother or sister. It makes the genetic constellation in each egg and each sperm unique.

Mendel's laws describe the rules of precedence that apply to gene variants and the results they produce, although only in the case of characteristics that arise from a single gene. Hair colour is one such characteristic. If we assume for the sake of convenience that just two colours exist, fair and dark, then there's a gene for hair colour that has two variants. Let's call them F for fair hair and D for dark hair. We all inherit either an F or a D allele from each of our parents.

If a baby boy inherits a D allele from both his father and his mother, then he is guaranteed to have dark hair. After all, there is nothing in his genes that could cause blondness. For the same reason, everyone with two F variants is fair-haired. But the assumption that children who receive an F allele from one parent but a D from the other will be dark blonde is wrong. They will all have dark hair, because the D allele is dominant. If a D variant coincides with an F variant, the D variant takes precedence. So the F variant is recessive.

Anyone with fair hair must therefore have the combination F-F, while anyone with dark hair may have inherited either D-D or D-F. This has consequences of its own. Although two fair-haired people will only ever have fair-haired children, since between them they have no D variant to pass on, two dark-haired people can have children some of whom are dark and others fair. If both parents are D-F and both pass on an F allele, the result is F-F and therefore fair-haired. Recessive alleles can therefore be handed down unnoticed for several generations before suddenly being expressed.

This may seem rather capricious, but the overall proportion of people with characteristics that are determined by single genes is in theory predictable. If both alleles are equally common, a quarter of the population will have two copies of the recessive variant, a quarter will have two copies of the dominant allele, and the rest will have one dominant and one recessive copy. This means that three-quarters of the population will display the dominant characteristic.

The occurrence of left-handedness cannot be explained based purely on the distinction between dominant and recessive alleles, since the number of left-handed people comes nowhere close to the 25 per cent that would be expected if a recessive allele were responsible. Still, variants do not always occur in equal numbers in a given population. In Scandinavia, for example, fair hair is so common that the F allele of the hair colour gene must be present at an extremely high rate in the population. Outside Europe, and leaving aside communities of people whose forebears emigrated from Europe, you have to search long and hard to find fair hair, since the F allele is extremely rare. Blonde Chinese people exist but there are very few of them indeed. So if the left-handedness allele is relatively uncommon, wouldn't the actual proportions of left- and right-handers automatically roll out of the mix?

This sounds like a neat solution, but in fact it is a red herring. Differences in frequency between gene variants are the cause of what we call racial characteristics: skin colour, hair colour, the shape and colour of the eyes, and other things of that sort. Even slight isolation in combination with different climatic conditions is sufficient to produce striking superficial differences between peoples over a period of roughly 10,000 years. And they are not confined to useful but otherwise innocent adjustments to features such as eye shape and hair colour. Sickle cell anaemia is a serious, recessive disorder that causes a defect in red blood corpuscles, making them sickle-shaped and unable to do their normal work properly. It's rare almost everywhere, but in areas where malaria is widespread the incidence of the condition is far higher than normal. The reason is that it offers protection against the even more damaging malaria parasite, which spends part of its life cycle in healthy red blood corpuscles and multiplies far more slowly in sickle cells.

The distribution of hand preference bears no resemblance to any of this. In all times and all places it's almost precisely the same. Hand preference is therefore a characteristic unaffected by circumstances; there seem to be no selective pressures. Yet sensitivity to selection is at the heart of Mendel's laws. He played God by crossing different strains and dictating the circumstances under which his plants were allowed to reproduce. The power available to us by this route is demonstrated by the simple fact that the Great Dane, the poodle, the beagle, the Pekinese and the naked chihuahua were all produced in the space of a few thousand years by selective breeding from the same ancestor. We have achieved astonishing variation – yet paw preference in animals has proven impossible to influence by selective breeding.

In a rather desperate attempt to rescue the idea that hand preference follows the pattern of Mendelian inheritance, it has been suggested that the recessive left-handedness allele of a presumed hand-preference gene works badly, so even people with two recessive alleles for left-handedness can end up right-handed. But this is thoroughly unconvincing. For one thing such a gene variant would have to work very badly indeed, given that left-handedness is more common than average in monozygotic twins but in the vast majority of cases affects only one of them. At the same time it must work well enough to continue to produce left-handers, as a hard and fast rule, at a rate of 10 per cent in any given human population.

There is one fact that definitively closes the door on a Mendelian account of the distribution of hand preference. Any explanation in terms of a recessive allele would inevitably mean that two left-handed parents could produce only children who were, genetically at least, left-handed, whereas in fact the majority of children of such parents are right-handed.

The best attempt yet to link left-handedness with genes was made around 1970 by Marian Annett, a psychologist at the University of Hull. It occurred to her that the distribution of hand preference in humans and paw preference in animals, although different, is patently similar. In animals the proportions suggest that paw preference emerges entirely by chance: one quarter are left-pawed, one quarter right-pawed and the rest undecided. The same groups exist in the human population, Annette argued, as long as you ignore those cases of left-handedness that are attributable to some kind of physical damage. Annett believed there was a group of left-handers amounting to around 4 per cent of the population, a group of right-handers that made up some two-thirds of the total and a group of mixed-handed people that accounted for the remaining 30 per cent. If you convert these proportions into a graph, then as with animals you get the bell-shaped curve that statisticians regard as a normal distribution, the difference being that the human curve has been budged a considerable distance towards the right. In other words, by far the larger proportion of the curve – and therefore of people – falls into the right-handed region, while most animals are around the neutral middle point and therefore somewhere in between left-pawed and right-pawed. These are animals that don't care which paw they use.

The shift to the right is simply a representation on a graph of the well-known lopsided distribution of left- and right-handers, and Annett believes it's caused by what she calls a right shift factor, the product of a gene that does not occur in everyone. Those who have it are guaranteed to be right-handed. If you don't have it, then it's a matter of chance and of circumstances whether you'll be left-handed, right-handed or neither, and to what degree.

To make the picture clearer we might want to regard the right shift factor as a gene with two variants. One of them, let's call it R, causes right-handedness, while the other, which we'll call NIL, does nothing at all. NIL

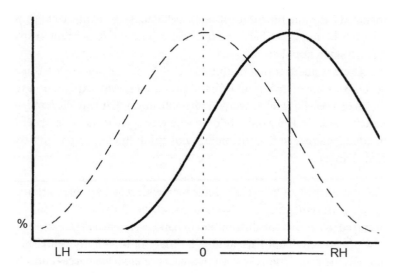

The distribution of left-handedness, right-handedness and mixed-handedness in animals (broken line) and humans, according to Marian Annett.

behaves recessively, Annett says, so everyone with an R-R combination is right-handed, as is everyone with an R-NIL combination. Only individuals with NIL-NIL may perhaps by left-handed, depending on other conditions. If those conditions are entirely neutral, then again we would expect a chance distribution: one quarter of NIL-NIL people are left-handed, half have no hand preference and the rest are right-handed.

This provided Annett with a neat explanation not only for the uneven distribution of hand preference but for the fact that left-handers have a greater likelihood than right-handers of producing left-handed children. So far so good.

There is a further piece of evidence that supports her idea, arising from *situs inversus*, that rare reversal of the inner organs. It's a disorder that seems to be caused in exactly the way Annett describes. Research, some of it carried out on split salamander eggs, shows that *situs inversus* does not result from a genetic defect, as had been assumed, but from the absence of any instruction as to which way round the organs should be constructed. Roughly half of those salamander individuals that lack the instruction go on to develop normally, the rest have their inner organs reversed. So mechanisms of the type Annett describes do in fact exist.

Yet Annett gets nowhere when it comes to twins. Her theory doesn't help to explain the elevated number of left-handers among both monozygotic and dizygotic twins. For the former you might want to attempt to solve the problem by assuming that they include far more NIL-NIL individuals than the rest of the population, but this is a blind alley. After all, the brothers and sisters of twins have the same parents, so they too should have an elevated chance of a NIL-NIL combination, giving them as great a likelihood of being left-handed as the twins themselves. Almost all studies show this is not the case. And elevated rates in dizygotic twins remain a total mystery.

It seemed at this point as if every attempt to explain left-handedness as a product of genes was doomed to run into the sand. Moreover, a separate story was needed to explain cases in which left-handedness is accompanied by all kinds of ailments that point to brain damage. Many people were beginning to suspect that left-handedness might in all cases be the result of accident or injury.

30

Hormonal Left-handedness

Although we now know for certain that the Roswell incident of 1947 was a fairytale, in 1974, on a top secret military training ground in the American state of Nevada, a UFO really was shot down by the US Air Force and held at a secret location. One of its occupants lived for some weeks and was interrogated at length by America's most secret agents. Wherever secrets are held they will leak out and so it comes about that we know the UFO came from Mars, and that it was not on its first mission. We also know how profoundly Martians differ from us. They reproduce without sex, a process from which they emerge as adults. They do not have gender, they do not understand the concept 'child', and they are unwavering rationalists – which is how they came to master interplanetary travel so soon, while human beings were still making their first paltry efforts at it.

The surviving Martian, consistently referred to in the secret interrogation reports as Alien 47 – an irresistibly evocative fact to conspiracy theorists – said that this was their first mission to earth and that their surface transporter had been shot out of the sky right at the end of their mission. They had just put back on earth the last abducted human to have been studied onboard the mother ship, which had been lurking all that time unnoticed on the dark side of the moon. Over the course of a three-month stay close to the earth, they kidnapped fifty to sixty people and laid them on their examination table, hoping to solve a puzzle that had held the Red Planet in its grip ever since the autumn of 1971 (earth years).

On 27 November that year Mars was struck for the first time by an object originating from earth. It was the Russian Mars 2 orbiter and lander, which was sadly dashed to pieces. The Martians managed to save only a small Russian flag from the wreckage, on a stand of chrome-plated steel wire – the interrogators had to rack their brains for a long time before they worked out this was what Alien 47 was trying to describe, since the Martian had no earthly idea what flags were. Five earth days later the Mars 3 flew into the planet's thin atmosphere and made a successful landing. The shocked Martians lost no time in getting hold of the contraption and rendering it harmless, which is why the Russians lost radio contact almost immediately. This time they recovered from the craft a plaque with what the Martians saw as rather primitive mathematical symbols. But there was also something that intrigued them immensely. It was a pile of passport photos that a Russian technician had slipped into the spaceship to give aliens an idea of what human beings looked like.

It seemed from the photographs that there were two kinds of earthling, a very common sort with a smooth chin and a rare kind with what could be described as shrubbery around the mouth. The Martians, who were all as alike as peas in a pod, had never seen anything of the sort before. They decided to mount a mission to the earth to find out what bizarre life form they were dealing with. This was the mission at the end of which Alien 47 fell into earthmen's hands.

During the Martians' study of the abducted and anaesthetized humans, it turned out that on some smooth chins there were fresh little cuts, as well as red spots on the adam's apple. No matter how carefully they searched, those cuts were never found hidden in amongst chin-shrubbery. Alien 47 said proudly that the specialists on board, after puzzling long and hard, worked out that all the smooth chins must have been created by an endlessly repeated ritual lopping process. Clearly that daily trauma did not usually affect people so much that visible marks were left, but in some cases it did. Therefore, they concluded with relief, there was only one kind of human after all, just as there was only one kind of Martian.

The interrogation transcripts, typed out verbatim, show in vivid detail how the astonished interrogators, sometimes unable to suppress a chuckle or two, tried to disabuse Alien 47 by telling him the human population was made up of men and women. And that there was something called childhood. And that children never had beards and therefore, like women, didn't need to shave. Which was why there were so many perfectly smooth chins. But Alien 47 went on stubbornly shaking his head. That couldn't be true, he insisted. Femaleness, childhood, a theory involving such bizarre notions made no sense to him, especially when there was a far simpler and perfectly adequate explanation. They must think he was stupid, he reproached the agents. Imagine, he said *mockingly, if he were to come home with such an unconvincing story. As soon as the governors of Mars had stopped laughing they would deport the whole crew to the feared mines of Baf! Then, exhausted, Alien 47 fell into some kind of coma from which he never awoke. What eventually became of him, nobody knows.*

Of course Alien 47 was wrong, but he did exactly what every sensible scientist is supposed to do: he looked for the simplest possible explanation for the phenomena he observed. After all, the fewer assumptions you rely upon, the fewer flawed assumptions you can make. So the stubborn fact that some cases of left-handedness are attributable to brain damage caused quite a few people to believe that all left-handedness might have its origins in trauma. In other words, people concluded that left-handedness is generally a result of damage to the brain, which affects control of the true preferred hand to such a degree that the other hand functions slightly better, yet is not so severe that any other effects can be readily observed.

Even though many of the studies linking left-handedness to some kind of disorder are of dubious quality, there are so many of them that we have to acknowledge that left-handedness can sometimes be a result of disturbed development in the womb or of brain damage caused before, during or shortly after birth. Aside from associations with all kinds of ailments and disabilities, connections have also been found between the occurrence of left-handedness and long or difficult labour, caesarean section, a relatively high age of the mother, incompatibility of the rhesus factor and other comparable risk elevators. Less than satisfying though it may be, we are forced to conclude that almost everything that can possibly go wrong while a fertilized egg develops into a newborn child may have left-handedness as its main result, or as a side effect.

This does indeed make it extremely tempting to see pathological left-handedness, the result of physical trauma, as the only kind there is. It's tempting in both its simplicity – it leaves us with a single cause, whereas all other theories inevitably start off with at least two forms of left-handedness – and in its concreteness and plausibility. There's nothing strange or difficult about the idea of damage to a specific bit of brain tissue, and of course damage does sometimes occur in the womb or during labour and its aftermath.

Another advantage of a general theory of trauma is that it makes the raised percentages of left-handedness in twins easy to explain. The two foetuses are crammed tightly together, which in itself involves certain risks, and one of the two always has to wait longer during childbirth. This entails an increased risk of oxygen deprivation and, who knows, damage that causes left-handedness.

It sounds convincing, yet the foundations on which this kind of reasoning is based are far from sound. First of all, trauma theories depend on the unspoken and unproven assumption that we are all right-handed by nature. This of course cannot be taken for granted, so proponents of this theory, like their rivals, have to explain two phenomena. Which will not be easy, since research on animals shows they have no general preference for either the right or the left.

Second, the simple fact that 10 per cent of the population is left-handed would mean that at least 10 per cent of people start life with an injured brain. In fact the implications are far worse still. In all groups in which left-handers are over represented, the overwhelming majority are nevertheless right-handed. The number of people going through life with brain defects must therefore be huge if this theory is correct, since if the left-handedness of someone who suffers from a particular disorder points to a problem with their brain, why would that not also be the case with at least some of the right-handers in the same group?

Conversely, left-handers are vastly more common than the ailments with which they are associated. Alien 47 saw many more unblemished smooth chins than cut and scraped ones, and he wrongly attempted to explain this based on a single cause: shaving. That's precisely what's happening here. The trauma theory means that practically every knock an unborn child receives leads immediately to left-handedness and only occasionally to other abnormalities. This seems extremely improbable.

Nor should we forget that far from all associations between left-handedness and a disorder of some kind necessarily point to trauma. A trauma is something that does not occur naturally in an organism but is suffered at some point during the course of its life because of external influences, or as the result of some kind of defect. Take for example the Dutch research carried out in 2005–9 that discovered a connection between breast cancer in young people – which is fortunately quite rare – and left-handedness. If there is indeed a connection, this would point towards a non-traumatic factor that by pure chance increases the odds both of left-handedness and of susceptibility to early breast cancer. It might well be a variant of a gene, but it could also have something to do with the regulation of hormones.

The existence of pathological left-handedness is a far less compelling motive to look to trauma for an explanation than it appears to be at first sight, because in the absence of a good reason for thinking that we are all right-handed by nature, pathological right-handedness must exist as well. The thing is, we just fail to notice it.

The invisibility of pathological right-handedness is a direct consequence of the uneven distribution of left- and right-handedness and the fifty-fifty chance of a trauma occurring on either side of the brain. Try the following simple calculation.

We start with 10,000 children, of whom 10 per cent are naturally left-handed. For simplicity's sake we'll assume that brain traumas that cause a reversal of hand preference occur in one child in every hundred. Then the following happens:

- In the course of their development 50 children will suffer damage to the left side of their brains and 50 to the right.
- Of those with damage to the left brain, the natural right-handers become left-handed as a result. There are 45 of them. No change in hand preference occurs in the other five, the natural left-handers.
- Of those with damage to the right brain, the natural left-handers become right-handed as a result. There are five of them. No change in hand preference occurs in the other 45, the natural right-handers.
- At the end of the process the number of left-handers is 1,000 - 5 + 45 = 1,040, including 45 pathological left-handers. This means that almost 1 in 20 left-handers is a traumatic left-hander.
- At the end of the process the number of right-handers is 9,000 - 45 + 5 = 8,860, including five pathological right-handers. So only 1 in roughly 2,000 right-handers is a pathological right-hander.

The influence of trauma on the size and composition of the group of left-handed children is significant, but the tiny handful of pathological right-handers is lost in the ocean of right-handedness of which it is part. If the group of left-handers increases by 4 per cent, the proportion of left-handers in the population as a whole increases by around half of 1 per cent. This fits nicely with the experience that in groups with some kind of disorder the number of left-handers is often slightly larger than normal. But even to explain such a small increase we have to assume that in one in a hundred children a trauma has occurred that affects the preferred hand just enough to make the other hand do its work instead, while at the same time producing effects so slight that no one notices anything wrong. After all, everyone is under the impression they're dealing with a normal left-hander.

These are stringent demands to make of traumas, events that have a natural tendency to affect people in random places to a random degree in random ways. If that one specific type of trauma that influences hand preference occurs so often, how improbably high must be the total number of accidents large and small? Or is there a risk factor that often causes this specific type of trauma? One man who has travelled a long way down that particular road is Norman Geschwind.

From 1969 until his death in 1984 at the age of only 58, Norman Geschwind was professor of neurology at the renowned Harvard Medical School. He was one of those strapping men who achieve an almost godlike status in their field, the kind of man who always thinks a little further, a little faster and better, coming up with more original ideas than other people. And he thought: testosterone!

To Geschwind the concept of trauma did not mean that someone who was left-handed would inevitably experience negative consequences. On the contrary, he believed that left-handedness was merely an easily detectable indication that the person in question might to a greater or lesser degree have a brain that was organized differently from the norm. He claimed this anomaly coincided with a number of other noticeable characteristics, bringing with it an increased risk of being a man, of having autoimmune diseases ranging from hay fever to the strangest of allergies, of getting migraine, being born with a hare lip, suffering from epilepsy or presenting with various other aberrations. It might well be that these were offset by a decreased risk of a series of other disorders, much like the way that carriers of sickle cell anaemia are less susceptible to malaria. It sounds bizarre, but it does indeed turn out, for example, that schizophrenics with twin siblings generally have far better mental health when one or other twin is left-handed than when both individuals are right-handed.

Geschwind spotted a clue to the causes of left-handedness in two phenomena. The first was the fact that left-handedness occurs slightly less often in girls than in boys and the second was the strange and intriguing but fairly firmly established connection between left-handedness and autoimmune disease. To Geschwind the central question concerned the connection between these two facts. In some way or other, left-handedness had to do with the distribution of functions between the two halves of the brain. How autoimmune disease could be linked to that pattern of distribution was a complete mystery.

Geschwind began by tackling the simple question raised by all these mysterious findings: what is the main difference between girls and boys in their intrauterine development? The answer he found was testosterone, the male sex hormone. Boy embryos pretty much swim in it, producing it themselves, in contrast to girl embryos.

Boys are generally more susceptible to disorders of the immune system, so the next question was whether it might be possible to make a connection between autoimmune disease and testosterone. There too Geschwind succeeded. High doses of testosterone have an inhibiting effect on the thymus gland, which carries out important tasks that affect the development and precise regulation of the immune system. The next step was to make a connection between testosterone and the brain, where left-handedness originates. Here again the resourceful Geschwind came up trumps. It was already known that the left side of the brain starts to develop slightly earlier than the right, and now it turned out that high concentrations of testosterone reduce the speed at which the left brain develops, precisely in the period when that half is particularly vulnerable. This phase therefore lasts longer, which creates a dual risk: the left brain can more easily be damaged and there's a chance it will be overtaken and outflanked by the right brain, which begins to develop slightly later. The result, he concluded, would be an increased risk of disorders such as dyslexia – and of left-handedness.

His theory seemed tidy and watertight, but one question remained: where do left-handed girls come from? Although girl embryos don't make their own testosterone, girls are almost as likely to be left-handed as boys. And girls too suffer from the ailments that Geschwind linked to a disruption of the testosterone level in boys, if slightly less commonly. He found a solution in the fact that mothers also produce some testosterone as a by-product of all kinds of processes. This meant girl embryos could be exposed to a testosterone level that was higher than normal for them. So it was not so much a matter of exposure to testosterone as such, which differs a good deal between girls and boys, but of relatively high concentrations.

A cleverly constructed story, and there's no doubt a good deal of truth in it. It's also unique in that it's the only theory to make a credible connection between the slightly higher incidence of left-handedness in boys and their slightly greater susceptibility to all kinds of autoimmune diseases, allergies and so forth. But it can't be the whole story, because again it breaks down when it comes to monozygotic twins. How can we explain that when a monozygotic twin is left-handed, the other twin usually isn't?

It might be possible to explain this in the case of boys. Boy embryos, after all, produce their own testosterone, and two out of three pairs of twins share the chorion or outer embryonic membrane but not the amnion, the inner embryonic membrane. As long as the excess of testosterone produced by one of them stays within that inner membrane, the other will be unaffected. But why would one of a pair of identical twins overproduce testosterone and the other not? It can't be a result of genetic factors, since those are exactly the same in both cases. But then, so is the environment, because influences from outside the mother ought to affect both twins equally.

Geschwind has even more difficulty with twin girls. The testosterone that he suggests may be a cause of the various abnormalities in girls always originates in the mother's body. How two genetically identical unborn females could respond differently to it is a mystery.

31

How Even Detrimental Characteristics Can Survive

So that's the position we find ourselves in after a century of research. We're left pretty much empty-handed, since none of the three best explanations for the existence of left- and right-handedness is satisfactory.

General trauma theories have a long list of shortcomings, of which the most important is that they require us to assume that in newborns the world over, the incidence of brain damage is horrendously high, although we barely notice it in the course of people's lives. The 2009 discovery that differences between left- and right-handers in the arrangement of features in the two sides of their brains can be far greater and more general than was once thought makes it harder than ever to assume that some kind of trauma lies behind all left-handedness.

Geschwind gets us a long way with his testosterone theory, but it doesn't solve the twins problem. He's therefore unable to explain the existence of left-handedness as such, although the suggestion that testosterone is a driving force behind pathological switches of hand preference seems perfectly sensible.

So we're left with genetic explanations of the kind proposed by Marian Annett. When she presented her right shift factor it was no more than an abstract idea. Genetics has moved on since then and her factor has been fleshed out with some appealing biology. For example, it has been proposed that the shift is caused by a recessive gene variant for left-handedness on the x chromosome that has low penetration. The reason why it seems logical to expect to find the gene on the x chromosome is that it enables us to explain the slightly higher incidence of left-handedness in boys. Males have only one x chromosome, so its alleles are always dominant – there's no copy of any of its genes to stand in the way. In boys the recessive left-handedness allele could therefore take full effect, whereas in girls it would need to be present on both x chromosomes.

The concept of penetration is brought into play at this point because the effect of dominance as a result of the lack of a second x chromosome is far too strong. Boys are only slightly more likely than girls to be left-handed. Penetration is a refinement of the old Mendelian division into dominant and recessive gene variants. Over time it has turned out that dominant alleles are not always expressed in individuals in whom they are present. This is the case with polydactyly, for example, a condition in which a person has one or more fingers or toes too many. Among its causes is the possession of the dominant allele of a single gene, yet only two out of three carriers of that variant actually have additional digits. The allele is therefore said to have a penetration of 65 per cent. The main result of incomplete penetration is that dominant characteristics, like their recessive counterparts, may skip generations before suddenly re-emerging.

Penetration describes just how free a genetic variant is to exert its influence, but until it becomes clear how the brakes are applied this explains nothing. And since no one has determined conclusively which gene causes hand preference, such a theoretical allele is merely 'in the frame', to use police terminology – no more than a possible suspect against which there is as yet insufficient evidence to merit taking it in for questioning, let alone any hope of a conviction on grounds of convincing and legally admissible proof.

Even if we managed to adjust the penetration valve so cleverly that the presumed hand preference gene on the x chromosome came extremely close to producing the precise chances of inheriting left-handedness that exist in reality, we wouldn't really have got very far, because a gene variant can never directly explain the distribution of hand preference. The fate of Darwin's famous finches shows why.

In 1835, halfway through his famous journey on HMS *Beagle*, Charles Darwin caught and stuffed all manner of birds that lived on the Galápagos Islands, including a number of rather unsightly finches of various kinds that were extremely similar in appearance. The males were jet black, while the females had mousy, grey-brown plumage. The main visible difference between the species was the shape and size of their beaks. Only many years later, long after he returned home to Down House in Kent, did Darwin realize the deeper significance of this variation. Under the influence of the highly variable foods available on the different islands, a single species of finch had given rise to six separate, specialized species. Where the food on offer took the form

Three of the six species of finch that are found only on the Galápagos Islands. At the top left the small-beaked *Geospiza parvula*, at the top right *Geospiza fortis* and below them *Geospiza magnirostris*, whose name means 'with a large beak'.

of large, hard seeds, individual birds with big nutcracker beaks were at an advantage and became dominant, until eventually all the finches on those islands had large beaks. On islands where the birds were dependent on small seeds that were tricky to extract, the opposite happened.

We now know that the shape of a bird's beak is dependent upon two gene products, BMP4 and calmodulin. The more BMP4 an avian embryo produces, the broader and deeper its beak, whereas calmodulin makes a beak longer. The contrasting beak shapes of Darwin's finches represented the first of many genetic divergences. Over the longer term the finch populations on the islands, isolated from each other, built up such a large number of genetic differences that they became separate species.

Although we generally think of evolution as a slow process, nature can sometimes take rapid strides. In 1982 several individuals of the finch species *Geospiza magnirostris* were blown across to the Galápagos island of Daphne by a severe storm. On that island, where large, hard seeds were available, the coarse-beaked finches flourished, far more so in fact than their less well endowed native cousins *G. fortis*. But in 2004 drought hit Daphne. It was so severe that hardly any large seeds could be found that year. Only cactus seeds were available, and they were difficult to work loose from the plant using a large beak. The population of newcomer finches declined precipitately and the previously disadvantaged individuals with smaller beaks took the lead once more. *G. magnirostris* died out. Even among the native species, individuals with the smallest beaks quickly gained ground and within a short time after the drought the average size of the birds' beaks fell considerably. This meant not only that a new, different balance between species had arisen for the second time in a quarter of a century but that the characteristics of one species, under pressure of circumstances, had changed with what in evolutionary terms was lightning speed.

This is of course a very different matter from the emergence of a truly new species, but all that's important here is that genetic variation instantly came into play among the finches when conditions altered. It's a familiar pattern. Variation within its package of genes makes a species flexible and robust, increasing the likelihood that part of the population will be able to survive changes to the environment. What if all species of *Geospiza* on Daphne had been endowed with large nutcracker-style beaks? There would probably be no finches left at all on that island today.

The source of genetic variation is the creation and in some cases disappearance of alleles as a result of mutation. This is the motor of evolution: without mutation no new variation; without variation no selection, no adaptation to changing circumstances and eventually no living creatures. This means, however, that the relationship between carriers of different alleles is inherently unstable. The entire system, after all, is geared to making change possible. As a consequence of that instability, characteristics that aid successful reproduction will eventually spread through the whole population. Equally inevitably, everything that is disadvantageous will be selected against and eventually disappear. Since some genes are recessive, rare or have low penetration, this can sometimes take many generations, but truly detrimental inherited characteristics will not take long to become all but eradicated.

The extent to which gene-linked characteristics will be found that have no effect at all on reproductive success is unpredictable. Such traits ought really to cease to exist because there is no force ensuring their spread. The fact that they do sometimes occur is a result of what's known as genetic drift, the aimless floating around in the genome of neutral alleles. It happens as follows.

The intractable law that says favourable alleles spread through an entire population while harmful alleles disappear is a statistical one. It holds true for large populations as a whole, as measured over many generations. We can imagine such a population as an ocean of successive generations of individuals. Seen from an aeroplane, high above cloud level, the ocean looks like a broad, shining, virtually smooth surface. That is the average view. But from a ship on the water the sea may be calm and flat as a mirror, or an inferno of heaving waves higher than a house, or a washboard of raging white crests. In such large systems as seas and worldwide populations, considerable deviation from the norm can occur, usually at limited times and locations.

One such possible deviation is the spread by chance, within a small subpopulation, of a gene variant that has no effect. From there it can penetrate further into the population. This is what we mean by genetic drift. It's sometimes been claimed, precisely because the percentage of left-handers is so constant, that left-handedness has become anchored in the population by such a route. This is nonsense. The fact that a characteristic that has no effect neither becomes more common nor disappears as a result of selective pressure does not make it stable. It can disappear as easily and as much by chance as it appeared. That could happen here and not two communities away, or vice versa.

Instead of stability, one would expect characteristics that came into being through genetic drift to cause particularly large differences between subpopulations. Suppose for instance that a ship full of colonists of whom only 2 per cent happened to be left-handed arrived at a remote island and the travellers settled there. If left-handedness originally arose because of genetic drift, we would expect in theory that in the absence of selective pressure that percentage would remain constant. Many centuries later left-handedness would still occur in only about 2 per cent of the population of the island. Either that or the rare phenomenon of left-handedness would have disappeared altogether as a result of chance. This is not what happens.

If the characteristic distribution of left- and right-handedness cannot be deduced from existing, recognized biological mechanisms, then it is sensible to turn the matter on its head and begin by defining the characteristics of its distribution before looking for a mechanism that fits.

For all the mystery, there are one or two things that we do know for certain. We can be sure, for example, that there is an inherited aspect to hand preference and that the basic probability of its manifestation is an unchanging 10 per cent or thereabouts. We also know that left-handedness in a parent doubles the likelihood that a child will be left-handed. Oddly, every characteristic that conforms to these stipulations turns out to be self-stabilizing, as long as the product of the basic probability of its occurring multiplied by the influence of both parents remains below 100 per cent. Beyond that, the characteristic will spread unstoppably across all individuals.

Take our ship full of colonists on a remote island again. When they start to reproduce they form pairs that will sometimes consist of two right-handers, sometimes of a right-hander and a left-hander and sometimes of two left-handers. How many pairs of each composition there will be depends on the size of the group and the percentage of left-handers on board. But if every child that results has a 10 per cent chance of being left-handed and if that probability doubles with each left-handed parent, then after a while the colony will consistently produce left-handedness at a rate of about one in ten, no matter how many left-handers arrived with the ship and irrespective of whether the group grows, shrinks or maintains its original numbers. It doesn't even make any difference if there was only one left-handed person on the original ship, or only one right-hander, or 50 per cent of each. In all cases, after a handful of generations, 12.7 per cent of the population will be left-handed, a percentage that from then on will not change at all.

It barely matters at all whether a trait of this type is advantageous or disadvantageous in evolutionary terms. Which is nice, because one of the puzzling aspects of the existence of left-handedness alongside right-handedness has always been that no one could think what positive effect on reproductive success might ensure that left-handedness continued to exist. If hand preference were a matter of the emergence of gene variation under selective pressure, then there would have to be a positive effect of some kind. It might perhaps be something like the advantage of having sickle cell anaemia in regions where malaria is prevalent, but no one has ever been able to think of anything concrete. Now that we know that left- and right-handedness cannot be dependent on a duo of alleles, we are liberated from the requirement to come up with this elusive advantage. Evolutionary quality is of hardly any significance in the case of self-stabilizing traits.

To understand this we need to look at another human characteristic, one that has nothing to do with hand preference but does exhibit the same kind of stable distribution. That characteristic is sexual orientation.

population	% left-handers	couples			children	
		lh-lh	lh-rh	rh-rh	lh	rh
10000	**2,0**	2	196	4802	1040	8960
10000	10,4	54	932	4014	1219	8781
10000	12,2	74	1070	3856	1259	8741
10000	12,6	79	1100	3821	1268	8732
10000	12,7	80	1107	3813	1270	8730
10000	12,7	81	1108	3811	1270	8730
10000	12,7	81	1108	3811	1270	8730

population	% left-handers	couples			children	
		lh-lh	lh-rh	rh-rh	lh	rh
10000	**98,0**	4802	196	2	3920	6079
10000	39,2	769	2383	1848	1938	8062
10000	19,4	188	1562	3250	1425	8575
10000	14,2	102	1222	3676	1305	8695
10000	13,0	85	1135	3780	1278	8722
10000	12,8	82	1115	3803	1272	8728
10000	12,7	81	1110	3809	1271	8729
10000	12,7	81	1109	3810	1270	8730

How left-handedness stabilizes in an initial population of 10,000 people,
of whom in the first case 2% and in the second case 98% are left-handed, given
that the basic likelihood of left-handedness is 10% and each left-handed parent
doubles that likelihood. The average number of children per couple is two,
so the population neither declines nor increases. Within five or six generations
the stable distribution of 12.7% to 87.3% is reached.

Like hand preference, sexual preference can be divided, roughly speaking, into two categories, in the same ratio of one to nine. The majority of us are attracted to the opposite sex, while 10 per cent of us prefer people of the same sex. As with hand preference, the division is not razor-sharp and absolute, but most people are preponderantly one way or the other. It seems as if in sexual orientation there are more cases of explicitly mixed preference, but appearances may be deceptive. There are for example people who label themselves bisexual for ideological reasons. As with the number of left-handers, the percentage of homosexuals, male and female, seems to have been more or less constant for many centuries, all over the world. There is one clear difference, however. While it remains an open question whether hand preference is linked to any evolutionary advantage or disadvantage, there can be no doubt that homosexuality is a textbook case of a trait that makes people less likely to reproduce.

A confirmed homosexual has no interest in intercourse with the opposite sex and therefore in Darwinian terms he or she fails to compete. In fact from the point of view of both natural selection and sexual selection, a homosexual lifestyle is so disastrous that it's mystifying how this trait ever became established and why the numbers have remained constant even in the face of outright persecution.

We need to note at this point that the repression of homosexuality may actually have contributed to its survival. Throughout history homosexuality has been viewed with almost universal contempt, or indeed simply denied and oppressed. It was at best something engaged in by men at the margins of society. We need only look at the way people in strictly sexually segregated societies today turn a blind eye to homosexual contact as a safety valve for frustrations among young men. All men, homosexual or not, were expected to start a family and care for their children. Women were given no choice; they were simply presented with some equivalent of the Victorian adage 'close your eyes and think of England'. In many societies, sadly, they still are.

So for millennia most homosexual men and women had little chance of finding fulfilment, if we assume this would have involved sticking to relationships that suited their sexual orientation. They were forced to engage in family formation and reproduction to a more or less normal degree. The situation is probably little different today, not only in countries where homosexuality officially does not exist but even in Western Europe and America. So the negative effect on reproductive success has probably been considerably reduced by social pressures. It's far from negligible all the same. Which of us doesn't know at least one confirmed bachelor, not to mention a middle-aged man who has never flown the parental coop, house-sharing sisters, elderly spinsters or even a nun or a hermit? These are all lifestyles that have traditionally provided unsuspected, if perhaps unfulfilled homosexuals with a way to live.

As a self-stabilizing trait like left-handedness, homosexuality could easily acquire a place for itself within the human population. In fact as soon as it existed in a few individuals, for whatever reason, it would inevitably acquire such a place. The negative effect on reproductive success does not alter this, it only means that a slightly higher average number of children need to be born, across the board, to maintain the population level.

Take a concrete example. We begin with the same values as those for left-handedness, in other words a basic probability of homosexuality in any given child of roughly one in ten, double that if one of the parents is homosexual. We then assume that all pairs made up of two homosexuals fail to reproduce at all and that half of all couples of which one partner is homosexual are also childless. These are strict conditions, and they reduce reproductive chances considerably, but even then the average number of children per couple needed to maintain the population at its existing level only rises from two to 2.25. The number

population	% homosexuals	couples			children	
		ho-ho	ho-he	he-he	ho	he
10000	2,0	2	196	4802	1125	9900
11025	10,2	57	1010	4445	1227	9911
11138	11,0	68	1092	4409	1238	9911
11149	11,1	69	1100	4406	1239	9912
11151	11,1	69	1101	4405	1239	9912
11151	11,1	69	1101	4405	1239	9912

population	% homosexuals	couples			children	
		ho-ho	ho-he	he-he	ho	he
10000	98,0	4802	196	2	45	180
225	19,8	4	36	72	24	179
203	12,0	1	21	79	23	178
201	11,2	1	20	79	22	178
200	11,1	1	20	79	22	178
200	11,1	1	20	79	22	178

How homosexuality stabilizes in an initial population of 10,000 people, given that the basic likelihood of homosexuality is 10%, each homosexual parent doubles that likelihood, homosexual couples do not have children and only 50% of couples of which one partner is homosexual reproduce. On average each couple has 2.25 children. Whether the percentage of the population that is homosexual is 98 or only 2 at the start, within about five generations the percentage reaches 11. Room for variation lies only in the size of the resulting population.

of homosexuals in the population will remain absolutely steady at a little over 11 per cent.

It's more than possible that there are other traits of this kind, occurring at some constant frequency through all times and all circumstances, immune to selective pressures. We don't know of any, but that could be because they are uninteresting or inconspicuous, or occur invisibly, deep inside our bodies. They need not have anything in common, just as there is no connection between sexual orientation and hand preference, so they may have quite different causes and even rest on quite different principles. What those principles are is of course a mystery, but in the case of left-handedness we can have a go at identifying the driving force that lies behind the phenomenon, based on the remarkable patterns found in twins.

32

Left-handers as Undercover Twins

In 1967 the psychologist A. Subirana sighed that it almost seemed as if left-handed people were created purely to sour the lives of neurologists, but we might be equally justified in saying that twins were invented to sabotage explanations of left-handedness. Whatever approach we take, there they are, blocking the path.

Still, maybe there's a lesson here, an indication that we've always looked at twins the wrong way. Twins are a special group, so special that we're used to seeing them as the exception to the rule. What if that's not in fact true when it comes to left-handedness? What if twins, however rare, are actually the norm and there's something special about the left-handed product of a single birth? This feels like skating on thin ice, but it's an alluring idea nonetheless. Especially if for the time being we limit ourselves to looking at monozygotic twins.

There are at least three facts we can take as a starting point. The first is that left-handedness is almost twice as common among twins. This in itself is old news, but it becomes interesting when we combine it with a second fact: left-handed people seem to be roughly twice as likely to produce twin offspring as their right-handed brothers and sisters. This means there's a link both between twins and left-handedness and between being the parent of twins and left-handedness.

The third fact is that a left-handed person is roughly twice as likely to have a left-handed child as a right-handed person. If we take the liberty of combining this with the second fact, then we see that the chances of left-handers having twins and of having left-handed children are greater than normal to roughly the same degree. This looks promising, especially if we take into account that the incidence of monozygotic twins as a proportion of births varies hardly at all from one community to another, just like the proportion of left-handers.

Now we have to go one step further and make a bold assumption. Suppose not just that there's a connection between twins and left-handedness but that being a monozygotic twin is a precondition of being left-handed, such that only embryos that divide at an early stage can develop into either one or two left-handed foetuses.

The appeal of this idea lies in the fact that we then need only one characteristic to be passed down in some way from generation to generation in order to explain both non-pathological left-handedness and the existence of twins. It's clear what such a trait would have to involve: a stronger or weaker tendency for the early embryo to split. If a mother's body naturally exerts a constant pressure on that initial clump of embryonic cells to divide, and if that pressure is greater in the case of left-handed parents, then we come satisfyingly close to what our experience with twins and left-handed people already teaches us. We've found ourselves a theoretical trait that in every relevant sense causes hand preference and has some features of an inherited characteristic without being directly dependent on the presence of specific gene variants, which would have to occur in a precise, stable proportion of the population. But first we need to find our way around a couple of obstacles.

Even before we're properly out of the starting blocks with this new theory we come up against the incontrovertible fact that twins are far less common than left-handers. One in ten births produces a left-handed child, but monozygotic twins arise in only three or four cases in every thousand. To see how we can reconcile these figures, we first have to look more closely at what the creation of twins actually involves.

Monozygotic twins start out as such in the first eight or nine days after conception. It's then that a little bundle of cells, just starting to develop into an embryo, can sometimes split into two roughly equal parts, each of which is capable of growing into a complete and independent human being. That early embryo, although minuscule, is already fairly complex. The popular image of the fertilized ovum that immediately splits into two parts, roughly in the way an ordinary cell divides, is incorrect. That image arose out of famous experiments with salamander eggs, which when cut with a hair can develop into two new eggs, producing two viable individuals. Human twins generally arise at a much later stage.

On the fourth day of pregnancy a membrane known as the chorion grows around the clump of cells. In two out of three cases of monozygotic twins it surrounds both embryos, so clearly some two thirds of twins develop after this stage. Twins that have already split by this point each develop their own chorion. Around the seventh day another membrane develops inside the chorion, known as the amnion. This is hardly ever shared by twins, so by that time the split must have taken place. The majority of twins, therefore, must develop between the fourth and the seventh day after conception. The rare cases in which twins separate so late that they share the amnion as well as the chorion often involve unpleasant outcomes. They include all conjoined twins.

This relatively late splitting of cells means that the formation of twins is a fairly complex and risky process. By comparison, when a single cell splits little can go wrong. Division is something that comes naturally to most cells, but a multi-cellular structure is not easy to split in such a way that both parts fulfil all the requirements for further development. It is therefore an open question how Mother Nature gets away with it now and again.

Let's answer that question in the simplest way possible, by assuming that no special procedure is involved: if an embryo splits, it does so in a random manner. In the vast majority of cases one part will have insufficient material to develop further. It's a formless clump of a few dozen cells at most, so it can easily disappear without trace. This may seem inefficient, but it has the beauty of simplicity and it takes us a long way in the right direction, since it means that most children who at the beginning of pregnancy are one half of a pair of twins come into the world as single births.

You might think it ought to be easy to find out whether there's an element of truth in this idea, but that is not so. We're talking about an event in the first few days of a normal pregnancy. Future parents usually have no idea at this stage that they have conceived a child (leaving aside the various artificial ways of creating a pregnancy, which aren't of any help to us here). Moreover, even if we could get there in time to observe the course of events, what mother-to-be would be willing to make herself and her tiny embryo available for research from which neither she nor her child have anything to gain? It's no wonder that despite all the modern technology at our disposal, the development of monozygotic twins is still shrouded in mystery.

Time to go back to our starting point: left-handedness arises through the splitting of an embryo. A further indication that we are on the right track is that this splitting occurs precisely at the stage when the foundations are laid for the symmetries and asymmetries of the developing baby. An early embryo may be no more than a slimy bundle of cells, but it is rapidly developing the fundamental distinctions between back and belly, bottom and head. It could hardly be otherwise, since within about three weeks of conception the basic structure of a little human being is in place, with a tiny spinal cord, the very beginnings of different kinds of organs and muscles, a primitive system of blood circulation and even a heart. This is the tadpole stage in which the embryo is about one centimetre long.

Occasionally an inconsequential asymmetry in one of a pair of monozygotic twins will occur in the other as well, but in reverse. A well-known example is the shape of the ear, which on one side of a twin sometimes has precisely the same contours as the ear on the other side of the other twin. Further examples of mirror-imaging of this kind are eyelid shape and scalp hair-whorl direction. This could be a result of disruption to the calm, normal course of events, caused by the splitting of an embryo just when the first structural rudiments are being laid down. Thinking along the same lines, we might hypothesize that we are all theoretically destined to be right-handed but there's a chance that splitting will disturb an initial impulse towards the formation of asymmetrical features in the brain, so that one of the two embryos becomes left-handed, or even both.

It's important to remember that this can be no more than a slight disruption to the normal development of asymmetries if splitting is to have a happy outcome at all. We're dealing with a minuscule, extremely

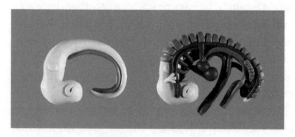

A model of a human embryo at roughly three weeks old, made by Colin Quilter of the University of Auckland in New Zealand. On the left the model shows only the neural tube that will form the spinal column, with the nodule at one end that will develop into the brain. On the inner side of the tube is the notochord, a primitive spinal column. On the right is the complete model, with the blood circulation and a tiny heart (the balloon-like structures at the centre). Across the back are the somites, segments of tissue from which the skin, vertebrae and muscles will develop. This stage is reached within fourteen days of the last possible moment when viable twins can be produced.

fragile scrap of tissue. Fortunately, if the process of division proves fatal, most would-be parents will never know the pregnancy existed.

We should not be surprised that significant reversals are rare, nor that where they do occur, as in the case of partial or complete *situs inversus*, they can cause serious problems. In the brain, which at the time of splitting has barely begun to develop, nothing too strange can be allowed to happen if the embryo is to survive. So the brains of left-handers usually look barely any different from those of right-handers and the most important functions remain in their usual places. In a minority some degree of reversal takes place, but as far as we can tell by the means now available to us, fMRI in particular, the most common effect seems to be that the brains of left-handed people are slightly less lateralized. Intuitively at least, this fits rather well with the idea that a minor ripple in early development is responsible, caused by the splitting of the embryo.

Based on what little we know with reasonable certainty, we can have a fair crack at calculating how great a tendency embryos have to split. The chance of left-handedness in any given individual is around 10 per cent. With twins the likelihood that both will be right-handed is about 64 per cent, while the likelihood that one will be right-handed and one left-handed is roughly 32 per cent. Anyone who's good at arithmetic can easily check all the sums, but for now we'll put the precise percentages to one side to avoid being distracted from the overall picture. Suffice it to say that they coincide neatly with the result that pops up again and again in all manner of research: left-handedness occurs in roughly one third of pairs of twins. The likelihood that at least one twin will be left-handed is slightly higher, at 36 per cent, since that figure includes the rare cases where both twins are left-handed.

It has also been shown that there's a chance of around three or four in a thousand that conception will result in a viable pair of monozygotic twins. Let's call it four. If 100,000 conceptions take place, they will produce 100,400 individuals, since 400 embryos will have successfully doubled to become twins. One in ten individuals are left-handed, so these 100,000 conceptions will produce 10,040 left-handers. Assuming they were all the result of the splitting of an embryo, we can calculate how many such splitting events must have occurred as a minimum. That number is equal to the number of left-handers divided by the likelihood of a pair of twins being produced with at least one left-handed individual. Dividing that result again by the total number of 100,000 conceptions gives us the likelihood that an embryo will split. It comes to a little under 28 per cent.

Since we know that out of every thousand conceptions only four actually produce twins, we now also know the likelihood that the splitting of an embryo will produce two viable individuals: a little less than 1.5 per cent.

Now we can work out how parental characteristics influence the tendency of their embryos to split. Twins are so rare that we can afford to ignore the influence of twin parents, but not of parental hand preference. Any embryo taken at random has an 81 per cent likelihood of having been conceived by two right-handers. Eighteen per cent have parents with different hand preferences, and 1 per cent are conceived by two left-handers. If we combine this data with the fact that each left-handed parent doubles the likelihood of splitting and twin formation, then we can conclude that with a right-handed couple, 23 per cent of embryos split, with mixed couples 46 per cent and with couples composed of two left-handers no fewer than 92 per cent.

Taking into account that the initial data are only approximate – the percentage of left-handers, for example, is not precisely ten; the chances of a left-handed child being born to a left-handed parent are not precisely double the norm – we can say that with right-handers a quarter of embryos split and with mixed pairs half, while the embryos of completely left-handed couples almost all split. This pattern must be the same when it comes to the chances of twins being born to parents one or both of whom is a twin.

So there we are. By this route we get exactly the right number of left-handers and the right number of monozygotic twins, along with an explanation for the fact that the frequency with which they occur in all population groups is stable. All this is based on just one assumption, namely that both phenomena are caused by the splitting of embryos at an early stage. The rest of the facts I have drawn upon here are well known from experience and research. Moreover, in this theory hand preference arises from a mechanism that we know causes other reversals, rather than from a special gene variant assumed to function purely as a factor that causes a switch in hand preference.

Nevertheless, we cannot prove this is how left-handedness arises. We'll have to leave that to the embryologists.

There are a few fuzzy edges left. It's no simple matter to explain how parental influence works. If the tendency to split is generated by signals from the mother's body, how can the father contribute to it? Further research is needed into the mechanism involved. Entirely left-handed couples are relatively rare, so impressions may be misleading. There's also a need to examine whether in the case of mixed couples it indeed makes no difference which parent is left-handed. As far as I know, there have been no systematic studies on the subject. It would be far from surprising if the mother's influence turned out to be greater.

A second loose end is the question of why it should be that we are all essentially right-handed by nature and not left-handed. The best explanation on offer is William Calvin's story about the reassuring sound of a mother's heartbeat, which is not altogether satisfying.

Finally there are dizygotic twins, who also have increased rates of left-handedness. We cannot solve that one, but perhaps we can lay it aside as less than wholly relevant, since dizygotic twins are far more common in some population groups than others. This suggests there may be no direct connection between the occurrence of non-identical twins and the occurrence of left-handedness. A small proportion of the increase in the likelihood of left-handedness can be attributed to trauma. Twins are squeezed together in the womb and often born prematurely or by artificial means. If they are not delivered by caesarean section then one of them always has to wait its turn, with all the attendant risks, of oxygen deprivation for instance. Who knows, perhaps the fact that over the past few decades a great many dizygotic twins have been conceived with the aid of technical interventions of one kind or another is a relevant factor.

As far as the causes of left-handedness are concerned, the ball is back in the researchers' court. Until we have more data about the theory posited here, there's little more to be said. It's therefore high time to look at the consequences of left-handedness, in all their many shapes and forms. Some are real, some imagined, some pure fantasy, and they range from the innocent or merely odd to the thoroughly disadvantageous.

33

The Consequences: Contrary, Perverse and Sick

'Gene for left-handedness is found', ran a BBC News headline on 31 July 2007. It went on: 'The Oxford University-led team believe carrying the gene may also slightly raise the risk of developing psychotic mental illness such as schizophrenia.' But left-handers shouldn't worry, the leader of the study reassured the public, since 'the vast majority of left-handers will never develop a problem'. Just how large that majority was, the news item did not say.

The ease with which such announcements are made is bizarre. No soup manufacturer would ever dare to put out a press release saying 'unfortunately, small splinters of glass have been found here and there in our tins of soup, but don't worry, the vast majority of consumers won't even notice'. Journalists would make mincemeat of the soup company boss, as would customers, food safety inspectors, politicians and the courts.

The same year further alarming reports emerged, this time from Australia, where a large-scale study had shown left-handed toddlers to perform considerably less well in every respect than their right-handed contemporaries. They also spent less time on educational activities and far more watching television. Scores were even lower for children without any clear hand preference. They were said to suffer from 'hemispheric indecision'. The researchers conceded that this last finding might arise in part from the fact that some pre-school children do not yet show a definitive preference for either hand, but they were nevertheless happy to declare that neither left-handedness nor mixed-handedness augur well.

At the same time they made mention of what might be called the most curious piece of research of the early twenty-first century, carried out at the renowned Johns Hopkins University in Baltimore, which concluded that highly educated left-handed men in America earn an astonishing 15 per cent more than their right-handed colleagues. Their enviable advantage was predictably hard to explain. It remains unclear, the American researchers said, 'whether left-handed graduates earn more because they perform better on the labour market or because they more often choose a more ambitious course of study and are more successful at it'. Either way, their findings were incompatible with the alarming developmental lag found in those 5,000 Australian toddlers. Falling behind and underperforming won't set you on the path to becoming an overachiever.

Equally strange is the fact that no one has detected such divergent outcomes anywhere else. If left-handers really do perform so poorly and have so many problems, then special education establishments, child welfare bureaus and youth detention centres ought to be overflowing with them. Which they are not. If left-handers really do earn so much more on average, wouldn't an aggrieved right-handed majority have rebelled against this injustice long ago? That hasn't happened either. In fact in everyday life no one seems to have realized how extraordinary and even dangerous left-handers are.

Fortunately, left-handed people don't usually pay much attention to dramatic announcements of this kind. They've grown accustomed to a great deal along these lines, most of it heard once and never again. Of course if three left-handers meet in a bar – a likelihood of one in a thousand, so it doesn't happen to them every day of the week – then they may perhaps briefly wallow in resentment towards the short-sighted educationalists who unnecessarily blighted their childhoods, but that usually exhausts the topic. Left-handers have no desire to adopt all-out victimhood. Shops selling left-handed items usually struggle to stay afloat – the British online store Anything Left-Handed being the exception that proves the rule – and most attempts to organize seem destined to fail. Only the United States turns out to be large enough to sustain a Lefthanders' Association.

This is not to deny that left-handed people frequently find themselves confronted with the consequences of their aberrant hand preference, but their difficulties are generally of a quite different order from those the right-handed imagine to exist. Moreover, consequences are not always disadvantages but may sometimes work in the left-hander's favour, and where problems do arise they're often less serious and easier to resolve than the average right-hander might think. Right-handers often watch left-handers the way an audience watches an illusionist sawing a lady in half: you don't need to be told that you shouldn't try this at home. The left-hander, like the illusionist, is a specialist in his field. Nature has given him more talent in his left hand than any right-hander can boast, and since early childhood he's found ways of dealing with a world full of right-handed people. The lasting impact of an outside world that from time to time thwarts him and the greater flexibility he's forced to adopt seem, at least, to cancel each other out.

Although we have plenty of preconceived notions about left-handers, there's no such thing as serious 'leftism' – or should that be 'rightism'? – and prejudice against left-handers has few practical consequences outside school. Nevertheless, it's infuriating, even downright dangerous, that scientists of fame and repute heedlessly contribute to their perpetuation. They make it extremely tricky to find out whether left-handedness really does have physical or mental consequences, as well as perpetuating the risk that people will wrongly be seen as suffering from some kind of disorder, or even written off by society as somehow inadequate. Abram Blau tried exactly that, as did Cyril Burt and more recently the Canadian scourge of left-handers Stanley Coren. Early psychoanalysts too, such as Wilhelm Stekel and Wilhelm Fliess, followers of Sigmund Freud who loved playing around with symbols and old popular maxims, laid themselves open to much criticism on these grounds.

In the early years of the twentieth century Stekel claimed that in dreams 'left' symbolized crime and therefore indicated homosexuality, incest and perversion. Fliess went even further: he transposed Stekel's dream theory to the real world and added to it the ancient connection in folklore between the left and the feminine. 'Where there is talk of left-handedness,' he pontificated, 'the disposition that belongs to the other sex is more in evidence. Not only is this verdict always correct, the reverse is also the case: when a woman seems like a man or a man like a woman, we see an emphasis on the left side of the body. Now that we realize this, we have the magic wand in our hands that allows us to discover left-handedness. This diagnosis is always correct.' Even that wasn't enough for him, since the combination of what he regarded as male and female qualities amounted in his view to degeneracy. It therefore didn't surprise him at all that there were 'so many' left-handers to be found among criminals and prostitutes. He went ahead and published without making the slightest effort to verify what he was saying, and without giving any thought to the consequences this kind of piffle might have for left-handed people in the sexually repressed society of his time – and indeed the effect on homosexuals of yet again being lumped together with criminals. Freud was enthusiastic about his good friend Fleiss's ideas about bisexuality and the combination of male and female sexual dispositions, but to judge from a personal letter he sent in response, he was less charmed by talk of a connection with any particular side of the body. Nevertheless he did not feel it necessary, then or later, to distance himself publicly from Fliess's pronouncements.

The worst thing about pseudo-scientific fantasies of this kind is that they are put forward by gentlemen of repute who present themselves as scientists, as people making a sincere effort to gather knowledge that everyone can rely upon. After all, that's what being a scientist is all about.

Most people involved in scientific endeavour don't act irresponsibly, but the influence of thoughtless clichés and deeply rooted mythology remains appreciable. This is demonstrated most plainly of all by the ubiquitous tendency to regard left-handedness as a problem. Disproportionate attention is paid to any association between it and specific ailments and disorders, while other possible side-effects of left-handedness are ignored. This has resulted in a gradual emergence of the idea that every left-handed person must have something wrong with them, even though on closer inspection there's little basis for any such conclusion. Of course there are left-handed people who suffer from dyslexia, hay fever, stuttering, personality disorders, retardation or whatever you care to name. Among people with some of these disorders we do find more left-handers than would be expected based on their share of the population alone, but the connections are weak and the overwhelming majority of left-handed people suffer from none of these things. We don't know whether there are any links between left-handedness and characteristics that in no way constitute a handicap. With a handful of exceptions, no one has ever tried to find out.

Research that focuses lopsidedly on associations with health risks creates an unjustifiable impression that left-handedness points to a poor state of health. There are no grounds for any such conclusion, as Norman Geschwind demonstrated by way of the following example. Women in the modern Western world run a considerable risk of a range of serious illnesses that do not occur in men. Think for example of cervical or ovarian cancer. Pregnancy and childbirth are also reserved exclusively for women and despite all the care available nowadays they are still not risk-free. Yet women of all ages have lower mortality rates than their male contemporaries, because the risks they face are more than counterbalanced by a reduced likelihood of other, often fatal afflictions like heart attacks and lung cancer.

When it comes to left-handedness specifically, the prognosis for schizophrenic twins is as clear-cut as it is puzzling. The sufferer will be considerably better off if at least one twin is left-handed, and it needn't even be the schizophrenic one.

There's as yet no reason to assume that left-handed people in general differ very much from the rest of us, either physically or mentally. If there is a difference, then it's too small for us to draw any conclusions about a person's state of health based on his or her left-handedness. That would be as absurd as to look at a man who is bald and conclude that he's being treated for cancer, or being poisoned with arsenic by his wife, even though both could certainly lead to hair loss.

Perhaps all we can say, intuitively, is that left-handers are on average somewhat less sociable and slightly more headstrong than normal, although we have no proof even of this. Those characteristics have no essential connection with the left hand, but left-handers learn even during their first attempts to tie their shoelaces that they will have to stand on their own two feet rather than simply copying everyone else. Left-handed people always have to do something extra, reversing the procedure demonstrated to them. Right-handed children never have to do this, except in those rare instances where both parents are left-handed. Even should they want to, most right-handed parents and teachers are simply not in a position to demonstrate a left-handed way of carrying out everyday tasks. They are generally content to be able to tie their laces and neckties at all – and who could blame them?

In any case, it's not at all easy to make yourself aware of all the things that can be done either left-handedly or right-handedly, since basic manual skills tend to come naturally to us. Only left-handed people notice the difference, consciously or unconsciously, and look for solutions of their own. They usually find them pretty quickly. The result is that they remain inconspicuous, so other people take little account of their unusual hand preference.

So we come full circle. From an early age left-handers are forced to be slightly more self-reliant than average. That won't do much to engender meekness in them.

34
Two Left Hands: The Ford Scale

The clumsiness of left-handers is proverbial. You only have to think of the appalling image painted by child psychologist Cyril Burt of little dolts blundering blindly through a china shop. On top of everything else, they're said to be unable to tell their right from their left. But is this an accurate picture? Or is it another piece of conventional wisdom that has more to do with the logic of our thinking in opposites than with concrete, demonstrable facts?

Firstly we need to decide what we mean by dexterity. One thing it certainly doesn't include is metaphorical nimbleness, in other words artfulness, a finely developed ability to slip through all kinds of legal, conventional and ethical nets. Nor do we mean sleight of hand, resourcefulness or ingenuity, an instinct for finding an unexpected but workable solution to every problem. It's true that excellent manual dexterity can help make us more resourceful, but the main requirements are analytical skills and creativity. The lightning fingers of the conjurer are not what we have in mind either, since he relies at least as much on psychological insight as on deft motor skills. A good conjurer plays more with his audience than with his cards. All these forms of dexterity make a person special in a positive sense, whereas a normal, inconspicuous person isn't expected to come up with any outstanding manual performances.

When we talk about dexterity in connection with hand preference, we are referring to how a person scores on the Ford scale. Vice-President Gerald Ford came to power in the United States more or less by accident in 1974, when Richard Nixon abandoned the sinking ship of his presidency. He remained in office until 1977 and he was a fantastic bungler, a man with an exceptional talent for tumbling down aircraft steps in public, bumping into waiters bearing full trays of drinks, and

President Gerald Ford in two characteristic poses: at the bottom of the steps to Air Force One and on the ski piste.

other embarrassing tricks of that kind. Ford was the sort of man who couldn't look at a book without contributing a smudge or a crease, who couldn't drink a cup of tea without leaving a footbath behind. He was a left-hander, and he proved both likeable and capable as a president, a rare combination. Yet he is remembered mainly for the fact that he had only a modicum of control over his large body – which was taken as incontrovertible proof of clumsiness in the left-handed – and so it seems appropriate to attach his name to the scale on which we measure this kind of control.

The Ford scale provides an assessment of performance in the fields in which Ford so noticeably failed: everyday motor skills, and the ability to manoeuvre your body through the environment such that no one even recalls you were there. Unlike other forms of dexterity, these are skills at which you can distinguish yourself only in a negative sense. A high scorer on the Ford scale is a perfectly ordinary, inconspicuous person, someone who can use tools and utensils without instantly damaging them, doesn't spill anything when filling cups and glasses, and after a little training could work perfectly successfully as a restaurant waiter. On the Ford scale everyone receives two scores: one for manual dexterity and hand-eye coordination, and one for the ability to navigate through the everyday world.

As far as manual dexterity and hand-eye coordination are concerned, all manner of research shows that there's barely any detectable difference between left-handed and right-handed people. Researchers have looked at how quickly test subjects can stick pegs in holes, trace over a drawing, or drum their fingers on the table. Almost everyone turns out to be able to do these things better and more quickly with their preferred hand than with the other, which won't greatly surprise any of us, but it also turns out that left-handed people can generally work as quickly and accurately with their left hands as right-handers can with their right. There's a significant difference only when we look only at the most clumsy. In that group the percentage of left-handers is slightly higher than normal, a finding that can be explained by the relatively large number of pathological left-handers, in other words natural right-handers who have become left-handed because of some kind of brain damage and are therefore in fact working with their non-preferred hand because with the other they're even less proficient.

As far as dexterity is concerned, therefore, left-handers will generally achieve normal scores on the Ford scale. Their reputation for clumsiness must be purely a matter of perception. They're regarded as clumsy no matter how well they perform, and this isn't so illogical as it may seem. Sometimes the cause lies in the design of instruments and appliances, which are not suitable for left-handed use, but there are other reasons that have nothing to do with the clumsiness of left-handers but everything to do with a lack of understanding on the part of the majority right-handed world for the minority left-handed world.

Left-handed people, for a start, do a lot of things the opposite way around, or at any rate differently from right-handed people. To the right-handed majority this always looks odd, even disturbing. The way a left-hander sets about knitting, slicing bread, or tying a necktie goes against all the rules deeply ingrained in every right-hander. One mother with a far from clumsy but nevertheless left-handed daughter was known to run out of the kitchen every time the girl started to slice bread, vegetables or meat, exclaiming that she couldn't bear to watch those frightening goings on. Kitchen knives are dangerous things and ever since childhood it had quite rightly been impressed upon the mother that they should be used with great care and precisely as taught. Irrespective of the impeccable results of her actions, her daughter clearly had a habit of going against the inviolable laws of knife-usage.

Second, left-handed people sometimes use methods that seem a little strange, or less than elegant. One important reason is that they have received incorrect instructions, or none at all. The countless little skills that have to be learnt before we reach puberty are demonstrated right-handedly nine times out of ten, which is fine for right-handed children. They only have to copy the procedure as closely as they can. This won't work for left-handers. They have to figure out for themselves how to reverse the process they're being called upon to learn, if that's possible at all in any straightforward way. Right-handers really ought to have more respect for the creativity of their left-handed brethren in practical matters, since they almost always come up with a perfectly good alternative, even if it defies right-handed rules.

On top of this comes the fact that the work of a left-hander often creates an awkward situation for the person who comes along next. Any right-hander allocated a desk that's been used by a left-hander will be dismayed to find that all kinds of things are on the wrong side. Anyone attempting to slice bread that a left-hander has been at first stands a good chance of having to tackle a loaf cut at a slant in the wrong direction. The left-hander, by contrast, is completely at home with reversals of this kind. Nine times out of ten he or she takes over the work of a right-hander. It will therefore seem perfectly normal that things on the table are the wrong way round, that the lamp illuminates the work of the right hand rather than the left and so on. It no longer occurs to him or her to complain about the mess that a 'clumsy' right-handed predecessor has made of things.

Now we come to that other score on the Ford scale: body control. Left-handers have a poor reputation here as well, as if they were people who stumble and blunder their way through life like gangly teenagers learning to manage their suddenly much longer limbs. This is the kind of performance the French refer to as gauche, literally left. Again there's probably little substance to the belief. No data have been collected on the subject – assuming it's possible to design a test for such a thing.

Yet the absence of facts and figures doesn't deter professional theorizers, people like the Canadian psychologist Stanley Coren, for example, who believes that left-handers are clumsy in social intercourse because they have a natural tendency to turn anticlockwise, whereas right-handers turn clockwise. As a result, the theory goes, they often bump into people.

Superficially this line of reasoning seems quite sensible. It's a more or less well-known fact that the majority of people arriving at a fork in a path or entering a room have a tendency to turn right, and therefore clockwise. Supermarkets and exhibitions are often laid out in such a way that by repeatedly turning right you automatically walk past the maximum possible number of articles or artworks. It's a tendency reflected in more formal situations too: all things being equal, if you're required to turn around, you're supposed to do so in a clockwise direction. In the armies of English-speaking countries the order to rotate through 180 degrees is simply 'about turn!' It apparently speaks for itself that you do this by turning to the right. In ballroom dancing most turns and especially the most complicated are to the right, which means that the leading man is required to turn on his axis and not around his out-stretched left arm. This undoubtedly helps him to keep his balance. According to Coren, left-handed people have an irresistible urge to turn left, so they go against the flow and cause a great deal of turbulence all around them. He takes left-handed Gerald Ford as his prime example. It has to be said: the man was a master at that kind of thing.

Coren even carried out an ingenious experiment that appeared to confirm his theory. Appeared to, because as we shall see his reasoning was sound but not the conclusion he drew. The experiment went as follows. In a bare, square, windowless room, with the one door in the middle of the back wall covered with a curtain, he placed two precisely identical tables to the left and right of the centre, each with an identical chair, so that seen from the door, the room was completely symmetrically furnished. Before entering, the research subjects were given a pen and a list of questions and asked to fill out the answers inside the room. The list included questions about the participants' hand preference, but in fact Coren was mainly interested in which of the two tables his subjects chose to use: would they turn left or right, and did their choice coincide with their hand preference?

The result was that the majority of subjects went to sit at the table to the right of the entrance to the room, just as they were expected to. They included more than two out of three right-handed people but only one in three of the left-handed subjects. The latter seemed to have a tendency to turn left that was 2.5 times greater than that of the right-handed. For Coren this proved the point: left-handers turn the wrong way, which explains their reputation as accident-prone.

But Coren spoke too soon, so pleased was he to have his prejudices confirmed. Look at what his results mean in the real world of a dance hall, for example. We would expect some 10 per cent of couples to be led by a left-handed man, the remaining 90 per cent by a right-hander. Based on Coren's figures, a left-handed male dancer will make 2.5 times as many mistakes on the turns as a right-handed one. So if a right-hander fouls things up once in an evening by turning the wrong way, then each left-hander will do so 2.5 times. If there are a hundred couples in the dance hall, then the right-handed among them will cause crushed toes or bruised shins ninety times, with the left-handers responsible for 25 collisions, ten times 2.5. This means the likelihood that you have a left-hander to thank for a mishap is only around 20 per cent. Eight out of ten times it's a right-handed klutz who wreaks havoc. That's hardly a reason to single out left-handers as a particular target for resentment.

At fault two times out of ten and yet given all the blame, that's surely rather unreasonable. Again it's the combination of conventional wisdom and the way in which some people perceive what they're seeing that causes left-handers to be regarded as clumsy, irrespective of the facts of the matter. If a right-hander treads on your toes, then he's an annoying, careless character, and that's an end to the matter. If a left-hander trips you up, then the reaction of people like Coren is to say: 'What did I tell you? They're all the same.'

Finally, popular wisdom attributes to left-handed people an inability to tell left from right. There may be some truth in this. Left-handers have learned since early childhood to reverse almost every activity demonstrated to them before copying it. This continues into adulthood: what right-handers do one way, they generally do the other. Worse still, what another person calls the right is often, for them, the left. It's at least conceivable that this leads to permanent uncertainty about left and right. But let's not forget Freud when indulging in this kind of speculation. Freud was a confirmed right-hander, who admitted to his friend Fliess not only that as a child he'd had difficulty with left and right but that even as an adult those paired concepts caused him problems. Spatial awareness was, he frankly admitted, not his strongest suit. Doubts of the Winnie the Pooh variety are clearly not the exclusive preserve of the left-handed.

35

The Things That Things Make Us Do

According to the ancient Greeks, at the beginning of time, before human beings came on the scene, the universe was ruled by the primeval god Uranos. His sons were the Titans, strapping lads of whom the youngest was Cronus. The rebellion of Cronus against Uranos marked the start of the everlasting rivalry between fathers and sons. It's a struggle that every son, when he eventually becomes a father, will at some point lose.

In those untainted, pre-worldly times, some less than subtle things happened. Take the way Cronus stripped his father of power. He took an adamantine sickle in his left hand, castrated his sleeping father and threw the sickle into the sea along with his father's crown jewels. Then the ambitious youth banished his unmanned begetter to the underworld for ever, taking his place on the throne of heaven until the day came when he in turn was violently dethroned by his son Zeus. Out of Uranos' crudely severed testicles and the sea, Aphrodite was born, the goddess of love. So Cronus' horrific deed brought forth something good in the end.

Like all myths, this story explores life's eternal themes, the instincts we all obey. What makes it unusual is the significance it attributes to the left hand. Sickles are agricultural instruments that cannot be switched from one hand to the other and have always been available exclusively in right-handed versions. Cronus' use of his left hand must therefore have symbolic significance. Maybe it was a way of emphasizing the unholy nature of his deed, a rebellion against the established order.

What goes for sickles goes for most other tools, appliances and machines: they are all made for use with the right hand, leading to thousands of minor problems for left-handers. Knives with one flat side are consistently sharpened in such a way that a left-hander will tend to cut on a slant. Scissors have loops in their handles that are the wrong way round for left-handers, causing painful rubbing, even blisters, and they are assembled the wrong way round as well. When considerable pressure needs to be applied, the blades, instead of being pushed together, are forced apart, so the material between them buckles – right-handers encounter this phenomenon when they try to cut the nails of their right hands. Measuring jugs with marks on the inside are impossible to read if used with the left hand; indeed, there's a whole series of ordinary kitchen utensils that are not so ordinary for left-handers, including corkscrews, saucepans and gravy spoons with pouring lips, and of course can openers. Then there are brooches and badges with pins that open the wrong way. A fish slice, fish knife or cake fork with a cutting edge is completely unusable for a left-hander, as are those modern plastic throw-away forks with serrations along one side. And rulers. The numbered marks run from left to right, forcing left-handed people to draw lines towards the zero, a source of irritating errors.

It's astonishing how many things are designed for right-handed use. The arm of an old-fashioned record player, for example: it's always on the right, as are the slots in coin-guzzling machines such as parking metres and juke boxes, or the grooves along which you swipe your debit card. The ignition lock on the steering column of a car, wherever you are in the world, is on the right, as are the most important knobs on audio equipment, such as the volume control. The buttons on monitors and television sets, if they're not at the bottom, are to the right of the screen. This seems innocent enough, until you realize that as a left-hander you have to reach across in front of the screen to adjust the colour contrast. Film and video cameras are shaped in such a way that they can be carried on the right shoulder only and have to be used with the right hand. A stills camera is operated by pressing a button that's always on the right.

There's more. Irons with those flex-holders that are supposed to be so handy are a complete disaster. Suturing materials used in operating theatres and first aid posts are designed for right-handers, not to mention the rest of the equipment. Left-handed seamstresses and tailors find that the controls on sewing machines are on the right, and that a great deal of discomfort arises from the fact that they insert pins the other way around. When those pins have to be taken out of the material while it's being stitched they inevitably have their heads towards the stitching foot, making them extremely hard to remove.

Electric drills have a blocking button, so that the trigger doesn't have to be held in all the time. It's always on the left side of the handle, which makes it completely useless for a left-hander – he switches that function off again with the palm of his hand. Downright dangerous are handheld circular saws, electric hedge trimmers and chainsaws. They can be used by left-handers only in a thoroughly irresponsible manner. Unfortunately the truly clumsy left-hander will be the one who fails to realize this. The situation is no better in industry. Factory machine tools and control panels always have buttons and knobs designed for use with the right hand. Sometimes even the emergency switch isn't replicated on the other side.

Particularly odd is the way our clothes are made. In our traditionally male-dominated world, men's clothing is buttoned and unbuttoned using mainly the right hand. Any man who breaks his right arm will discover that his generally so humble and cooperative flies have turned into an unmanageable, capricious monster. Women's clothing does up the other way. This must be a consequence of our deep-seated tendency to divide and polarize. It condemns the great majority of women to a lifetime of struggling to button their clothes.

This may explain one of those bizarre differences between men and women. Men always unbutton their shirts before taking them off. Women approach the task differently. They undo the absolute minimum number of buttons and then pull their blouses or cardigans over their heads. If you ask them why, they say it's too much trouble to undo and do up all those buttons. Ninety per cent of them are probably right: it's much harder with your non-preferred hand.

Leaving aside the fact that the numerals are on the right side of the keyboard, the digitalization of the world has brought a number of real blessings. In the paper era the counterfoils in chequebooks always presented a particularly annoying obstacle to left-handed people. The fewer cheques left over, the lower the writing surface fell below the pile of stubs on which the left-hander was forced to rest his or her hand. Occasionally a bank might be willing to provide reversed chequebooks on request.

Many counters and ticket offices have been replaced by websites. This has made life easier for left-handers in one way at least. We have largely been relieved of those pens on chains that organizations obligingly placed on their counters. The chains were always anchored to the right, so that when the pens were used by left-handers they stretched right across the form that needed filling out. And the chains were often so short as to make the entire operation impossible.

Another boon is the disappearance of telephones with the receiver attached by a flex. It was always fixed to the left side of the phone so that a right-handed person could easily dial or tap in a number and then have his writing hand free. That flex perpetually got in the way of left-handers.

Left-handed people always adjust amazingly well. They use their non-preferred hand much more than right-handers do and so become better at it. Left-handed implements are occasionally available, mainly in the kitchen goods and writing supplies sectors, but they're always hard to find. The only left-handed tools that are fairly readily on offer in the better ironmongeries are scissors.

There are good reasons why availability is extremely limited despite a theoretically enormous potential market of more than half a billion people worldwide. Generally speaking, by the time a left-hander discovers that a left-handed version of an article is available, he's already invented a strategy for dealing successfully with a right-handed model. Perhaps more importantly, anybody who gets used to working with a special left-handed tool will never be able to wield one belonging to someone else, the boss for instance, with the same degree of dexterity. So for a left-hander there's only a limited advantage to be had, with the exception of things that are used by just one person in one place, or are easy to carry around. In general a left-handed person will prefer to make do with the less than ideal right-handed version.

This usually works fine. With good quality scissors the paper or fabric that's being cut will buckle only if considerable pressure is applied. A reasonably practical solution is to use large, good-quality scissors for every cutting task. Only at nursery school, where children are given those awkward, round-tipped toy scissors, do left-handers find themselves completely stymied. A ruler, as we have seen, is perfectly usable for drawing a line of a specific length if you move your pencil towards the zero. That's not an ideal solution, but it does work. It's possible to compensate for the absence of left-handed versions of tools and utensils in a myriad of different ways.

There are certain circumstances in which making do is not acceptable, where nothing less than the optimum is good enough and money is no object. Architects, engineers and people trained in technical drawing, for example, traditionally used left-handed or right-handed drawing tables. Happily, equipment for dentists and dental hygienists is available in versions for left- and right-handers, and the same goes for the layout of operating theatres, which can be adjusted to suit the surgeon. In the medical world, much thought has been given to the hand preferences of those who probe, prick and cut us, and there have been debates about and research into the effects of left- and right-handed surgery on the success of specific operations.

Still, the ease with which left-handers make right-handed equipment their own and the care taken in the medical world do not alter the deplorable fact that most ergonomists and industrial designers have a thoroughly nonchalant attitude to the interests of left-handed people, even when the consequences can be dangerous. It must surely be possible to invent a neutrally designed or easily reversible hedge trimmer. Of course there's no apparent call for one; no left-hander would have the audacity even to inquire whether such a thing existed, since in his experience the answer has always been no. There will never be any demand without supply.

Incidentally, this lack of creative interest and economic incentives has nothing to do with discrimination against left-handed people. Designers and ergonomists only tinker at the edges, they never stop to think about the business of hand preference more generally, as became clear when the electronic till was introduced into supermarkets in about 1990. The cashier no longer had to key prices into a cash register; she only had to pass each article over the electronic eye that's become such a familiar feature of shopping.

It was a laborious affair at first. If anything it was slower than the old method, since many items had to be swiped across the eye several times before the machine would register the price. Cashiers thought at first that the eye must be dirty. Everywhere spray-bottles of glass cleaner appeared next to the till and cashiers sat there diligently polishing their electronic eyes. It hardly helped at all, since dirt was not the main problem. A lack of thought during the design process had produced an electronic till that was ideal for left-handers and unsuitable for 90 per cent of cashiers.

The old-fashioned supermarket till with its cash register was laid out so that the cashier sat with items on the conveyor belt arriving to her left. She picked up each article with her left hand and worked the till with her right. For left-handed people that was a trial, but it was the best conceivable arrangement for the right-handed majority: the left hand merely slid the purchases onwards, while the fine work was done by the preferred hand. With the arrival of the electronic eye the cashier's chair was turned through 90 degrees so that she sat facing the side of a surface set at the end of the moving belt, which now stretched away to her right, with a cash tray in front of her along with a keyboard

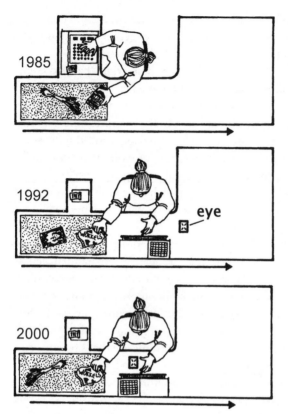

Comparison of supermarket tills with the electronic eye.

for entering the prices of unmarked items. The electronic eye was placed off to her left, since there, well past the end of the conveyor belt and the cashier's legs, was where most space was available.

This seemed logical, but it had unanticipated consequences. Suddenly the shopping no longer arrived at the cashier's left hand but at her right. That wasn't a problem in itself, but it meant that the tricky business of making the electronic eye respond had to be done on the other side, with her left hand. Most people were none too good at that part of the operation. Nowadays the eye is no longer positioned away to the left but always straight in front of the cashier.

36
Writing and Other Useful Handiwork

'Oh,' the eloquent sixty-year-old said over his beer. 'Yes, I'm left-handed, but at school they taught me to write with my right hand. And that was fine too. No problem.'

'But now you write with your left hand. You even drink with your left!' I slipped the beer mat on which he'd just scribbled his email address into my inside pocket.

'Yes . . . But at first I had to write with my right. So I just got on with it. Until I was thirteen. Then a teacher told me it was okay for me to write with my left hand if I wanted to, so I switched.'

'I see. So you're one of those people who were forced to write right-handed. Didn't it bother you? We're always being told that children start to wet the bed and so on, and stutter.'

'Nah . . . Not a problem. I can't remember anyone forcing me, really. I always enjoyed going to school. Yes . . . But I did have a terrible stutter, when I was young.'

'You? Nobody would think so now. You got over it all right!'

'Yes . . . It lasted until about the time I left middle school. Until I was thirteen, then it just went away of its own accord.'

'Until you were thirteen . . .'

'Yes.'

'When you started to write left-handed?'

'Around then, yes, I think so.'

'Might there be a connection? You started writing left-handed and promptly stopped stuttering?'

'Well I'll be damned. Never occurred to me. Wow. You know what – that could be it!'

Left-handers often come up with intriguing stories like this about their educational experiences, especially in primary school. As children they run up against a great deal of incomprehension and ignorance from teachers. They have to conform to a norm that's alien to them, or they're left to their own devices but continually told they don't come up to scratch. It seems they learn to overlook many things, the way this particular drinker never made a connection between his miraculously vanishing stutter and being forced to write with his right hand, and simply carry on undaunted.

Instead of waiting for help they look around for solutions on their own initiative, with such success that in no time they develop into perfectly normal pupils. Ask a teacher to divide a pile of homework from a school class he doesn't know by sex and he'll carry out the task whistling and with a fair degree of accuracy. Ask him to put work by left- and right-handers into separate piles and he won't know where to start. After a year or eighteen months left-handers are able to perform even that difficult trick of writing legibly, whether with their left hand or their right, just as well and as quickly as their right-handed classmates, despite the additional obstacles they face and the back-to-front education they receive.

Nonetheless it's unpleasant to have a teacher force you into all kinds of contorted postures your body is reluctant to adopt. You'd think the experience would inevitably leave its mark, but it's impossible to find solid evidence for the stuttering, bedwetting and other nervous disorders said to accompany being forced to write with the right hand. Still, such stories are heard so often that we can't simply dismiss them, especially when they're told by left-handers themselves.

Only a few decades ago it was still virtually universal practice to force schoolchildren to learn to write with the 'correct' hand. Enlightened teachers who allowed left-handed children to develop naturally, even perhaps managing to give them a couple of useful tips, were rare. Although excesses such as the strap or the tying of the left hand behind the back have been banned, the situation is not much better today. Even now in the modern Western world, children are all too often subjected to subtle pressure to give writing with the right hand a go. It would be better that way. Pedagogical wisdom has it that more than half of all left-handed children suffer from a secret ailment called 'habitual left-handedness' – they're not really left-handed, they're just pretending to be. The rascals! A characteristically right-handed form of concern about the delicate soul of the left-hander is common too; left-handed children are regarded as suffering from a sense of being misfits in a right-handed world. How treating them as children with special needs and an unfortunate natural preference could possibly help, goodness only knows. Another ineradicable myth is that our form of writing is so geared to the right hand that it's impossible for left-handers to master. Millions of left-handers furnish evidence to the contrary every day.

All these worries among educationalists are absolute nonsense, there's no other word for it. Anyone who has been left-handed since childhood knows that by the time a school pupil starts to learn to write his habits are firmly ingrained. Those in whom hand preference is not yet fixed but who eventually, despite all the right-handed models and examples they see, opt for the left hand are apparently not too bothered by the experience. Either that or they have quite different, serious problems. True, if you're left-handed you may sometimes be bullied as a result, but the same, or worse, goes for the colour of your hair, your accent, your name, the braces on your teeth, the shape of your nose, the brand of your clothing, your eccentric mother, your test scores and an infinite number of other things. No one ever goes to a social worker or a psychiatrist complaining of left-handedness.

The true reasons behind attempts to get left-handers to mend their ways have little to do with worries about their success in learning to write and far more to do with conformism and a misplaced desire for order. A left-hander disrupts the uniform image of the ideal classroom filled with children quietly beavering away, so he or she is forced to conform: all children must sit neatly two by two in rows, all with their arms folded, or all writing with the same hand. Left-handers create further insecurity by confronting teachers with their own incompetence, and teachers react by plunging their heads into the sand. They try to arrange things so that it seems as if the left-handed child doesn't exist. The more authoritarian the education system, the stronger this tendency towards denial. The more pressure there is to conform, the harder life is made for left-handers.

No one says this aloud, of course. Officially it's all about how the poor left-hander will benefit. The most popular argument presented for making children switch hands in learning to write is that dreadful hooking of the wrist. It's always assumed that left-handers make a smudged mess of their written work because they wipe the side of their hand through the wet ink, and because they can't see what they're writing because their fingers get in the way. This is assumed to lead to efforts to solve the problem by writing with a strange, crooked claw curled over the top of the line: the hooked hand. But this argument cuts no ice either.

Two variations on that dreadful hooked hand.

There are certainly a good many left-handers who write with a hooked wrist, often producing perfectly neat handwriting, but anyone who takes the trouble to look will see that a good many right-handers write in a similarly hooked manner. This seems bizarre, since it's impossible to imagine a reason why a hooked wrist would be comfortable or effective for anyone. Yet it is quite common. If you examine the way adults wield a pen you'll soon notice that there are ten or more ways in which people hold the thing. They include many that are almost painful to watch. Left, right and centre you'll see pens clamped between three or even four clenched-white fingertips, held between an index finger and a middle finger and resting high up against the palm, or grasped in hooked claws of every shape and description. Left-handers don't seem to adopt improbable writing postures any more often than right-handers do. There's only one possible conclusion: generally speaking the teaching of writing isn't what it ought to be, so children are often forced simply to invent approaches of their own, with variable results.

Another ineradicable story frequently heard in the teaching world is that so-called crossed lateral preference is a terrible thing. This is a belief that goes back to old-fashioned ideas about dominant halves of the brain. It used to be unquestioningly accepted that a healthy person's preferred hand, eye and foot were all controlled by the same cerebral hemisphere. Not only is there no trace of evidence for this, there's not even a clear connection between hand preference and other preferences, and oddly enough it's regarded as a problem only in the case of left-handed people. No one tries to convert right-handers into left-handers if they turn out to be left-footed or left-eyed.

Still, we shouldn't be surprised, because ultimately all this concern is purely for show. If a left-handed child can't be persuaded to write with his right hand, then in most cases the expert simply gives up and leaves him to his fate. It's a rare teacher who'll take the trouble to learn how to present the correct left-handed method. Left-handers are used to this, of course, since as with other activities like knitting or woodwork they have to puzzle out for themselves how to adjust the consistently incorrect model they're shown. Nothing can be relied on, other than meeting with contempt when they don't fully succeed straight away. Left-handers include few enthusiastic young seamstresses and craftsmen for precisely this reason.

The worst scenario of all is one in which a left-handed pupil is presented with a supposedly left-handed method of writing. It always comes down to teaching the poor left-hander a handwriting style that leans backwards, based on pseudo-scientific hocus pocus about natural movements and directions. This won't help the child at all, since the angle of lean is completely irrelevant. Worse still, writing that leans to the left is socially unacceptable. It makes you look odd. Any amount of neglect is better than being thrown from the frying pan into the fire by being made to stand out.

What all this amounts to is that left-handers in schools are still seen as problem cases. Even today teachers talk of 'tolerating left-handed writing', as if it were an undesirable habit. Left-handers are always treated as if they have a disorder, a shortcoming. They're seen as high-risk pupils, almost as if they should be approached only in the company of a school doctor or psychologist. If anything is damaging to the mental health of left-handers and their ability to enjoy life, then it's this.

Another deep-seated myth is that left-handers tend to produce mirror writing. At school this too is a source of misplaced concern. Outside school it's generally a straightforward misconception, but it can also be a trick that gains a left-hander a good deal of credit. It's quite common for left-handers to train themselves when young to produce mirror writing and to develop an ability to do so reasonably smoothly and neatly, some with the left hand, others with the right. It would be no surprise to find that right-handed people can learn just as easily to produce fast, neat mirror writing with their left hands, since this particular trick has nothing to do with the act of writing as such.

Much of the blame for the development of the mirror writing myth should be laid at the door of Leonardo da Vinci, who was left-handed and in the habit of writing backwards. We don't know why. That's a secret he took to the grave with him. It has been suggested it was a way

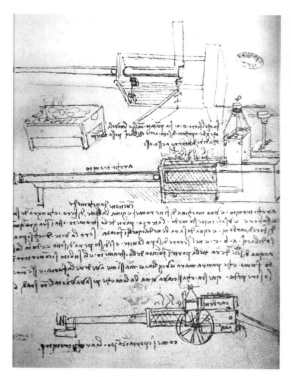

Leonardo da Vinci's design for a steam cannon made of copper, with notes in mirror writing.

of making his notes unreadable to spies and competitors, but this seems unlikely. His fifteenth- and sixteenth-century contemporaries weren't so stupid and naive as to be fooled that easily. Leonardo came from a family of lawyers and writing was therefore a prominent aspect of his life, which in a still predominantly illiterate world and in combination with his left-handedness may have been enough to persuade the intractable, restless young Leonardo to make mirror writing his own – for fun, because he could, to be different, and to amaze other people. Sigmund Freud, inevitably, thought it had to do with repressed sexuality. On 9 October 1898 he wrote to Fliess, his pupil at the time: 'Perhaps the most famous left-handed individual was Leonardo, who is not known to have had any love-affairs.' Hmm.

There was considerable fascination for mirror writing in the late Middle Ages, and people tended to confuse it with left-handedness. In 1540 Giovanni Battista Palatino in Rome published a course in calligraphy that became a bestseller. He devoted a chapter of it to *lettera mancina*, or left-handed writing. He was referring to mirror writing, which right-handed people can produce more readily using their left hands.

The confusion had set in well before then. Around 1560, sculptor and architect Raffaello da Montelupo, also a resident of the Eternal City, wrote in his autobiography about his experiences as a young left-hander 50 years before. He wrote:

> I will not omit to say that by nature I am left-handed, nor that, finding my left hand more facile than the right, I used to write with it, since my teacher made no objection, being satisfied that my handwriting was good. I have therefore always used my left hand, whether I was writing or copying pictures from the *Morgante*, which we read from at school. As soon as I laid the sheet lengthways to write with my left hand,* many people looked surprised; they thought I was writing *all'ebraica* [in the Hebrew manner, from right to left] and that what I wrote would turn out to be impossible to read. I remember one curious event in particular. When I started to write out a receipt for a certain amount for a notary in Florence, I laid the sheet lengthways and the notary expressed doubt as to whether what I wrote would be legible. As soon as I'd written one sentence he picked up the sheet of paper, saw that it was perfectly readable and called out to some ten other notaries to come and look. When the receipt was finished, I wrote a few words with my right hand, because I was well able to use that too, even though I'd stopped using it to write with as time went on. As I've said, I draw better with my left hand. Once, as I sat sketching the Arco di Trasi al Colosseo [the Arch of Constantine], Michelangelo and Sebastiano del Piombo came by and stopped to watch me. Now you should know that both of them, although left-handed by nature, did everything with their right hands unless they needed to apply force. They stood looking at me for a long time in amazement because, as far as is known, neither of the two ever created anything with their left hands.

As Da Montelupo delights in telling us, left-handed people do not produce mirror writing but handwriting like anyone else's. The only difference is that they write with their left hands. Without training, mirror writing is as difficult for a left-hander to produce as it is for the

---

* Da Montelupo probably means that he laid his paper crosswise and more or less wrote from top to bottom. It's a rather extreme version of the diagonal position of the paper you often see with left-handers, which we'll return to shortly.

average untrained right-hander. Should a child just beginning to write have an urge to start at the top right corner of the first page of his brand new exercise book, a simple cross next to the left margin to indicate the starting point is sufficient to set things straight.

Teachers are generally more concerned about reversed letters and words than about mirror writing. It's a familiar phenomenon. When learning to write, children initially have a tendency to write certain letters backwards, especially capital S and capital N, along with the pairs d and b, and q and p. But this is not specifically something that applies to left-handers. Even adults unused to writing often run into trouble in this way, as do people daubing words in unusual circumstances – as slogans in giant letters on bridges and walls testify. Of course it may be that left-handed children have a stronger and more lasting tendency to reverse letters than right-handed children, but as yet no evidence for this has been produced. Until it is, we're talking about nothing more than an impression, mainly in the minds of the right-handed majority. Generally speaking, such impressions have turned out to be less than reliable.

In reality it's not at all difficult to write well and comfortably with the left hand. Or rather, no more difficult than with the right hand. There are just a few things you need to know.

Anyone writing with their right hand does best to hold the pen fairly loosely between the thumb and index finger, resting it on the side of the middle finger. The act of writing consists mainly of two kinds of movement, a slight flexing of the wrist, which gives the up and down lines in the letters, and a slow sweep outwards from the elbow that allows the line to be filled in steadily from left to right. Roughly speaking, that sweep begins straight in front of the centre of the body. The paper is at a slight angle, with the top right corner higher, so that the underarm lies perpendicular to the writing when it's in the middle of the line. That way the writing hand remains under the line, where it doesn't get in the way of the writer's line of sight and minimal effort is required. If the wrist touches the page, it must not be leant on too hard if the script is to flow properly in its long sweep to the right.

Making upward lines thin and downward lines thick, as in calligraphy and elaborate, old-fashioned handwriting, is the logical way to stress how the letters are formed. On the downward stroke the wrist brings the hand towards the body, a movement that's easier, firmer and more controlled than that which forms the upward lines, away from the body. Right-handed writing has much in common with the wrist movement involved in peeling potatoes.

Left-handed writing happens in almost exactly the same way. Again we ensure that the writing arm is perpendicular to the direction of writing when the hand is in the middle of the line. So again the paper naturally lies at an angle, this time with the top left corner upwards. The way the pen is held is the same too: loosely between thumb and forefinger, resting on the side of the middle finger. The pen must be held such that when the fingers are at rest the tip is only about a millimetre from the paper.

Now too the writing hand is always below the line, so there's no need to smudge. The writer can easily see what he's just written. The long sweep from the elbow is once again the means of getting from one end of the line to the other, except that now it's a sweep not outwards but inwards, ending roughly in front of the centre of the body rather than starting there. So far, then, apart from the obvious reversal, everything proceeds in much the same way.

The really important difference in writing technique lies in the small movements that go to form the upstrokes and downstrokes. A right-hander uses an up and down motion of the wrist, whereas in a left-hander the direction of movement is at 90 degrees to this, roughly an extension of the hand and the pen. It involves simultaneously bending and stretching the thumb, forefinger and middle finger, which act in concert to direct the pen, a motion that closely resembles the removal of a splinter using tweezers.

Left-handed writing is a good deal more subtle and susceptible to changes in pressure than the potato peeling approach of right-handers. Inexperienced writers sometimes try to force their letters into shape by clutching the pen tightly and pressing it hard on the paper. This does nothing to improve even a right-hander's writing, but in a left-hander it's disastrous. A degree of attention and control is needed at the start, therefore, as well as a ban on the ballpoint, which only invites extra pressure. This will be more than enough to ensure a left-hander can write just as well as any right-hander. Even the 'upwards thin, downwards thick' comes naturally, since it emphasizes the inevitably light push to stretch upwards and the more powerful and confident pulling motion downwards.

It's true that some letters will be formed in a different way by left-handers, who discover the best approach for themselves. They may well have difficulty with school workbooks that show them how to trace out

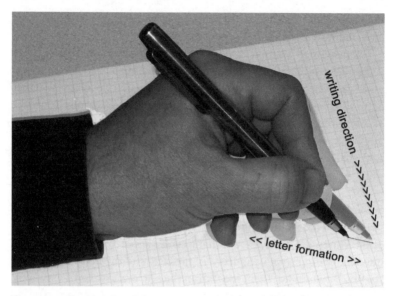

This is how good left-handed writing works. The paper is laid diagonally, so that the writing arm is parallel to the side of the paper and perpendicular to the line of text. The movement required to form the letters is not back and forth, as it is for right handers, but in and out, not unlike using tweezers to remove a splinter.

the letters, since left-handers are always presented with the wrong model. Any teacher who isn't too strict about how a left-handed pupil forms letters but concentrates instead on whether or not he's producing functional, legible handwriting will be doing him and his kind a great service.

37

The Myth of High Left-handed Mortality

Perhaps the most disturbing message ever delivered to left-handers came in 1991. Deploying his considerable skill at attracting publicity, Canadian psychologist Stanley Coren made it known that left-handed people die no fewer than nine years earlier than right-handed people. At that time right-handers in the United States were living to be 75 on average, whereas left-handers gave up the ghost at only 66.5. He was the first person ever to have looked at left-handedness and detected such disastrous consequences.

Left-handers, familiar with all kinds of fantastic fables about their idiosyncrasy, let even this onslaught wash over them with a sense of resignation. Not so Coren's fellow psychologists. Harsh criticism of the quality of the work of Coren and his faithful accomplice Diane Halpern soon appeared in professional journals, and even mainstream newspapers and magazines took note. The story nonetheless proved impossible to eradicate and it still pops up from time to time. There is therefore every reason to take another close look at why Coren's story is wrong.

Coren had spent years studying left-handedness when it struck him that as well as being more common in men than in women it occurred more frequently in young people than in the elderly. To judge by the answers to questionnaires, left-handedness disappears over the course of a lifetime. Or could it be, Coren asked himself, that it's the left-handers themselves who disappear? He decided to do some research.

First, in 1988, he carried out an exploratory study among American baseball players, the only group of people in the world whose hand preference is registered with painstaking precision. He concluded that left-handers die several months younger than comparable right-handers, a conclusion that researcher E. K. Wood of the California Institute of Technology demolished on technical grounds in *Nature* of 15 September

1988. As Wood convincingly showed, Coren had not done his statistical homework properly, so no conclusions at all could be drawn from his study.

Coren, however, continued undaunted along the path he'd decided to take. More extensive research was required, he decided. The main difficulty was to find data from a large number of more or less comparable dead people without shocking their families or coming up against overly scrupulous bureaucrats. Eventually, with the help of the local authorities, he succeeded in collecting the details of around a thousand people from a region of Southern California. He accumulated his data by asking families three things no sooner than nine months after a relative's death: which hand did the deceased use to write, draw and throw a ball? If the answer to all three was the right hand, then the person in question was regarded as right-handed, otherwise the individual was recorded as having been left-handed.

Here already, right at the start, it's clear that the research was seriously flawed. You may be able to say of a late member of your family which hand he or she wrote with, but are you equally certain about drawing? Even in the case of someone who wasn't in the habit of taking out a sketchbook? How often have you actually seen your father or uncle draw? Do you have any idea which hand your mother or aunt uses to throw a ball? How long ago is it since you saw them do such a thing? If you think you know, are you completely certain? People can be fairly unreliable when researchers ask them questions about their own hand preference, so serious doubts must surely arise about how much faith we can reasonably have in this kind of information about family members, especially when they've been dead for at least the best part of a year.

Coren went on to compare the ages at which his left- and right-handed subjects died. That was how he reached the shocking conclusion that the lives of left-handers were nine years shorter on average, and that at all ages they ran a greater risk of dying than right-handers from the same geographical region.

Of course it's true that even in the United States teachers used to treat left-handedness far more harshly than they do now, and that it was therefore the older members of the group Coren investigated who were most likely to have been intimidated and oppressed. Yet he was right to argue that this could not explain the enormous difference, since studies from the time when his subjects were young invariably showed that the percentage of people who openly stated they were left-handed was hardly any lower than it is now. Even in times of serious social repression, it seems people do not necessarily repudiate their left-handedness. Still, we need to be careful here. In those older studies people reported their own hand preference, whereas Coren's study was about someone else's. This is an important distinction.

Coren went looking for explanations for the astonishing outcome of his research, and he found them. In essence he believed that left-handed people had serious accidents far more often than right-handed people, sometimes with fatal results, and that long before they reached 70 a majority of them had died. He put this down to the fact that the world is arranged with right-handed people in mind, making it all the more dangerous that, or so he believed, left-handers are clumsier than their right-handed brothers and sisters.

One example of how he defended these conclusions is his study of car accidents. The psychologist claimed that left-handers cause more fatal crashes than right-handers do. He explained this by saying that left-handers have different reflexes. If someone is seriously startled, perhaps by having a ball thrown at his face, then a defensive reflex makes him put his hands up to protect himself, his favoured hand across his chest, in other words quite low down, and the other higher up, in front of his face. In a driver this same reflex, Coren claimed, meant that left-handers would tug the wheel to the left, into the oncoming traffic. Right-handers would tug it to the right, moving out of the traffic lane altogether.

It sounds credible, but in fact there's not a single grain of truth here. First of all an essential part of the shock reflex is that the hands are spread wide, with the palms facing forwards. In a car this wouldn't cause the horrified driver to tug at the wheel but to let go of it. If we don't take our hands off the steering wheel in an emergency, then clearly that defensive reflex simply doesn't occur, and there's no good reason to think that one particular element of it would.

Still, let's assume for a moment that Coren's shocked reaction is of some relevance. In that case the situation in countries where people drive on the left must be dire, with all those right-handers causing head-on collisions. Sure enough, Coren declares without batting an eyelid that the traffic in those countries is pure pandemonium, and he takes Britain and Ireland as examples. There, he says, the roads are incredibly unsafe, with on average more accidents than in the rest of Europe. True, no one had ever noticed this before. In actual fact Coren was comparing apples with oranges. To prove himself right he set the figures for British and Irish traffic accidents against the average for fourteen other European countries as a whole. If we seperately compare each of the countries Coren looked at with Britain, and with each other, then Britain and Ireland come out as run-of-the-mill average. Indeed research by the European Union, carried out in the first decade of the twenty-first century, identified Britain as having the safest traffic of all the 27 countries of the EU, with Ireland in a respectable seventh place. Heading the danger list are our old friends Spain and France, where people drive on the right, as they do in most of the rest of the world.

Third, Coren bases his argument that left-handers drive much more riskily on the total number of fatal accidents in which one of his deceased research subjects was at the wheel. He doesn't look at who was responsible. Worse still, he pays no attention to the nature of the crash, even though his reflex theory is relevant only to head-on collisions, which make up no more than a small proportion of the total number of accidents that result in fatalities.

Not only does Coren lump all road deaths together, the amount of data he draws upon is far too small. If we accept that American traffic is not hugely more dangerous than Dutch traffic, then based on the figures from the Dutch Central Bureau for Statistics relating to 1989, around 1 per cent of Coren's deceased will have been victims of traffic accidents. So we're talking about no more than ten or so fatalities, and those will include passengers, whose defensive reflexes are completely irrelevant. Still, let's give Coren every possible benefit of the doubt and accept that among his victims of crashes all ten were responsible for the calamity that struck on their final journey. Coren proposes that left-handers are involved in fatal accidents four times as often as right-handers. Assuming that the average percentage of left-handers is the traditional ten, which fits reasonably well with Coren's data, it turns out that Coren bases his sweeping conclusion on two or at most three accident-prone left-handers. In short, Stanley Coren is taking us for a ride.

One interesting and indeed fatal objection to Coren's data in general was contributed by Dutch science journalist Martin van der Laan to the daily newspaper *Trouw* of 9 March 1991. He began with Coren's claim that around 20 per cent of twenty-year-old men are left-handed but by the age of 50 this has declined to a mere 5 per cent. Coren claims the difference is explained by mortality rates. According to the Dutch Central Bureau for Statistics, around 5 per cent of all male citizens of the Netherlands die between the ages of twenty and 50. In other words, out of every 100 twenty-year-olds, an average of 95 reach the age of 50.

The figures are no doubt comparable in the United States. This makes it simply impossible for the percentage of left-handers to have fallen from 20 to 5 between those same ages purely as a result of a high death rate. Even if only left-handers died, at least 15 per cent of people would be left-handed at 50. Anyone still in any doubt can rest assured; Coren fell into the trap of his own enthusiasm, drawing nonsensical conclusions based on demonstrably unrepresentative figures and an unsound method of processing them.

Yet it remains true to say that left-handedness seems to decline fairly steadily as the age of the group questioned rises. Coren's basic error was to assume that this meant left-handers die young – a completely unfounded and far too drastic conclusion. If we ask groups of 50-year-old women whether they have recently become mothers, the answer will almost without exception be no, but we don't go on to conclude that fertile women die young.

The other cause most commonly put forward is social repression. The simple fact that left-handedness used to be treated more harshly is taken to mean that large numbers of people eventually forgot they were really left-handed. This sounds improbable, and so it is. In large-scale research in Britain in 1998 it turned out that including or leaving out questions about socially loaded subjects such as writing, and to a lesser degree drawing, had hardly any effect on the degree to which left-handedness declined in frequency as the age of the subjects rose. In 2000 more than a thousand people aged 65 to 100 were questioned, and no connection with social pressure emerged. The researchers were forced to conclude that 'age-related variations in hand preference can best be explained based on a number of factors of which the ways in which they influence each other is as yet poorly understood'. In other words: goodness only knows.

Yet ever since 1988 a far from implausible and fairly simple explanation for the fact that left-handers seem to dissolve into thin air has lurked within one of the studies Coren used to make his ill-considered bid for world fame. It's a study set up and implemented by psychologists Ellis, Ellis and Marshall using far better methods than Coren's and published in *Cortex*, a journal of neuropsychology. It looked at more than 6,000 men aged between fifteen and 70. In this particular group of subjects, the percentage of left-handers fell from just over 9 per cent among the youngest to just over 5 per cent among the oldest. Although less dramatic than the mass annihilation that Coren claimed to have discovered, this difference is far too great to be attributed purely to a reduction in social pressure on left-handers. Like Coren, the researchers based their findings on questionnaires, but theirs were far more thorough than his primitive list of three skills. People were asked to indicate, by putting a cross in the appropriate column, whether they had a preference for one or the other hand when carrying out each of ten different tasks. If they had a strong preference, then subjects could indicate this by entering two crosses. Each participant therefore responded with a maximum of twenty crosses and a minimum of ten.

At this point we need to ask how far the way in which the youngest and oldest distributed their crosses needs to differ in order to produce the reduction in left-handedness that emerges. The answer is rather surprising: depending on how many double crosses were entered on average, only one in every two or three of the subjects who answered the questionnaire had to move, all told, one cross from left to right. In concrete terms this means that if out of every two or three people there's one who in the course of a lifetime has switched hands for just one of the ten tasks, the difference is explained away entirely. In fact that person doesn't even need to switch hands. It's enough for him to have become less pronounced in his hand preference, so that instead of two crosses he enters one. Given the notion that we mellow with age, this is not at all improbable. Consider too the fact that the average fifteen-year-old is keen to be special and interesting. Left-handedness is out of the ordinary, so you'd be an odd kind of teenager if you didn't milk it for all it was worth. Once you reach 40, such considerations have come to matter a good deal less.

38
Creative, Musical, Brilliant and Famous!

'Is your child left-handed? Then maybe you're raising a genius', ran a headline in the British magazine *The Listener* in 1975. It was one of those articles in which enormous creativity and remarkable artistic gifts are attributed to left-handers, based on the supposed dominance of that emotional, creative right half of the brain. The value of ideas like this is of course only relative, even though they may come as a welcome relief to left-handers after a whole series of negative dictums, an attitude of incomprehension on the part of parents and teachers, and ominous messages à la Stanley Coren. It's often mentioned that not only the mentally handicapped but geniuses include an unexpectedly large number of left-handers. But what does this mean, if no one knows just how many geniuses the world has produced? People regularly point to famous left-handers of the past, such as Michelangelo and Leonardo da Vinci. In many cases, incidentally, this appeal to genius is based on shifting sand, since an enthusiasm for claiming prominent individuals for the left-hander camp often gets the better of conscientiousness. Bob Dylan is not left-handed, and neither was Picasso, nor a whole string of other celebrated figures. This kind of thing serves no purpose anyhow, since it would be equally justifiable and equally ill-conceived to comfort bald men who have birthmarks with the achievements of Mikhail Gorbachev, disabled people with the success of Stephen Hawking or depressive redheads with the genius of Vincent van Gogh.

Innumerable well-worn tales are told about architecture colleges and music academies where a fantastically high proportion of left-handers is to be found, and they're said to get better marks on average too. Anything is possible, but we have no solid data to suggest left-handedness is accompanied by special gifts. Only in two areas do left-handers undeniably have an advantage.

One is in sports like tennis, baseball, boxing and fencing. But then again, they're at a disadvantage in games like polo and hockey. The other is the US presidency. Bizarrely, that tiny, illustrious company of men includes an improbable number of left-handers. Of the last seven presidents, up to and including Barack Obama, five were left-handed, the exceptions being Jimmy Carter and George W. Bush. The Republican candidate for the 2008 election, John McCain, is also a member of the club, while in 1992 the sitting president, Bush senior, and both his challengers, Bill Clinton and Ross Perot, were left-handed. Still, does this say anything about left-handedness in general? If it does, then it can't be anything particularly complimentary, since nice people could never hope to triumph in the ruthless contest that is the race for the American presidency. We're talking here about such an exceptional group of people anyhow that it's safe to assume little can be deduced from the statistics.

Left-handers seem, as far as we can tell, simply to be left-handed. In a predominantly right-handed world they are generally presented with misleading examples to follow, unhelpful instructions and less than optimal tools. In many ways this throws them back on their own resources from a young age. Perhaps left-handed people should at least be allowed to flatter themselves with the thought that they're more independent than average.

Charlie Chaplin playing the cello with his left hand, *c.* 1915.

BIBLIOGRAPHY

Achbar, M., ed. 1995 *Manufacturing Consent: Noam Chomsky and the Media*, Montreal, Black Rose Books

Altera, J. 1970 'Wat doen we eigenlijk voor uitgesproken linkshandigen', *Pedagogische Studiën*

Anderson, M. G. 1989 'Lateral preference and longevity', *Nature*

Annett, M. 1972 'The distribution of manual asymmetry', *British Journal of Psychology*

——, 1974 'Handedness in children of two left-handed parents', *British Journal of Psychology*

——, 1975 'Hand preference and the laterality of cerebral speech', *Cortex*

——, 1976 'A coordination of hand preference and skill replicated', *British Journal of Psychology*

——, 1985 *Left, Right, Hand and Brain: The right shift theory*, London, Lawrence Erlbaum

——, and M. Manning 1989 'The disadvantages of dextrality for intelligence', *British Journal of Psychology*

Apuleius, L. 2003 *De gouden ezel (metamorfosen), vertaald en toegelicht door Vincent Hunink*, Amsterdam, Athenaeum – Polak & Van Gennep

Bächtold-Stäubli, H. 1987 *Handwörterbuch des deutschen Aberglaubes*, Berlin, De Gruyter

Bakan, P. 1973 'Handedness and alcoholism', *Perceptual and Motor Skills*

Bakan, P., G. Dibb and P. Reed 1973 'Handedness and birth stress', *Neuropsychologia*

Bambach, C. C. 2003 *Leonardo, Left-Handed Draftsman and Writer*, New York, The Metropolitan Museum of Art

Barnes, F. 1975 'Temperament, adaptability and left-handers', *New Scientist*

Barsley, M. 1968 *Linkshandigheid: wetenswaardigheden en curiosa voor links- en rechtshandigen*, Hilversum, Haan

Barsley, M. 1970 *Left-Handed Man in a Right-Handed World*, London, Pitman

Baskerville, R. F. 1992 'On the directional asymmetry of rhesus macaque forelimb bones', *American Journal of Physical Anthropology*

Bellugi, U., E. S. Klima et al. 1986 *Examining Language Dominance Through*

Hand Dominance, Boston University Conference on Language Development

Belmont, L., and H. S. Birch 1963 'Lateral Dominance and Right-Left Awareness in Normal Children', *Child Development*

Benbow, C. P. 1986 'Physiological Correllates of Extreme Intellectual Precocity', *Neuropsychologia*

Benton, A. L. 1959 *Right-Left Discrimination and Finger Localization*, New York, Hoeber-Harper

——, R. Meyers and G. Polder 1962 'Some aspects of handedness', *Psychiatric Neurology*

Best, C. T. 1985 *Hemispheric Function and Collaboration in the Child*, New York, Academic Press

Bever, T. G., and R. J. Chiarello 1974 'Cerebral dominance in musicians and nonmusicians', *Science*

Birkett, P. 1979 'Relationships among handedness, familial handedness, sex and ocular sighting-dominance', *Neuropsychologia*

Bisazza, A., et al. 1996 'Right-pawedness in toads', *Nature*

Bishop, D.V.M. 1983 'How sinister is sinistrality?' *Journal of the Royal College of Physicians of London*

——, 1989 'Does hand proficiency determine hand preference?' *British Journal of Psychology*

——, 1989 'On the futility of using familial sinistrality to subclassify handedness groups', *Cortex*

——, 1990 *Handedness and Developmental Disorder*, London, MacKeith

Blakeslee, T. R. 1980 *The Right Brain*, London, MacMillan

Blau, A. 1946 *The Master Hand: A study of right and left sidedness and its relation to laterality and language*, Research Monographs of the American Orthopsychiatric Association

——, 1961 'Don't let your child be a leftie!' *Lakeland Ledger,* 5 March

Bock, G. R., and J. Marsh, eds 1991 *Biological Asymmetry and Handedness*, New York, Wiley

Boklage, C. E. 1981 'On the distribution of non-right-handedness among twins and their families', *Acta geneticae medicae et gemellologiae*

——, 1984 'Twinning, handedness, and the biology of symmetry', in Geschwind and Galaburda 1984

Bouma, J. M. 1988 *Perceptual Asymmetries and Hemispheric Specialization*, diss., Vrije Universiteit Amsterdam

Bourassa, D. C., I. C. McManus and M. P. Bryden 1996 'Handedness and eye dominance: A meta-analysis of their relationship', *Laterality*

Bradshaw, J. L. 1980 'Right-hemisphere language: Familial and non-familial sinistrals, cognitive deficits and writing hand position in sinistrals, and concrete abstract, imageable-nonimageable dimensions in word recognition. A review of interrelated issues', *Brain and Language*

Bradshaw, J. L., and N. Nettleton 1983 *Human Cerebral Asymmetry*, Hemel Hempstead, Prentice Hall

Briggs, G. G., and R. D. Nebes 1976 'The effects of handedness, family history and sex on the performance of a dichotic listening task', *Neuropsychologia*

Brightwell, R. 1975 'Is your child left-handed? Then you may be bringing up a genius', *The Listener*, 7 February

Bryden, M. P. 1973 'Perceptual asymmetry in vision: Relation to handedness, eyedness and speech lateralization', *Cortex*

Burt, C. 1937 *The Backward Child*, University of London Press

Calnan, M., and K. Richardson 1976 'Developmental correlates of handedness in a national sample of 11-year-olds' (National Children's Bureau, London) *Annals of Human Biology*

Carlyle, T. 1837 *The French Revolution: A History in Three Volumes*, Boston, Charles C. Little and James Brown

Carroll, L. 1872 *Through the Looking Glass, and What Alice Found There*, London, MacMillan

Casasanto, D. 2009 'Embodiment of abstract concepts: good and bad in right- and left-handers', *Journal of Experimental Psychology*

——, R. Willems and P. Hagoort 2010 'Body-specific representations of action verbs: evidence from fMRI in right- and left-handers', *Psychological Science*

Castillo, G. A. 1974 *Left-Handed Teaching: Lessons in affective education*, New York, Praeger

CBS 1986 'Linkshandigheid', *Maandbericht Gezondheidsstatistiek*

Chamberlain, H. D. 1928 'The inheritance of left-handedness', *Journal of Heredity*

Clark, K. 1939 *Leonardo da Vinci*, London, Penguin

Clark, M. M. 1957 *Left-Handedness: Laterality characteristics and their educational implications*, University of London Press

——, 1980 *Linkshandige kinderen: onderzoek en praktijk*, Nijkerk, Intro

Cole, J. 1955 'Paw preference in cats related to hand preference in animals and man', *Journal of Comparative and Physiological Psychology*

Collins, R. L. 1968/69 'On the heredity of handedness', *Journal of Heredity*

Conrad, K. 1949 'Über aphasischen Sprachstörungen bei hirnverletzten Linkshänder', *Der Nervenarzt*

Corballis, M. C. 1983 *Human Laterality*, New York, Academic Press

——, and I. L. Beale 1971 'On telling left from right', *Scientific American*

——, and ——, 1976 *The Psychology of Left and Right*, New Jersey, L. Erlbaum

Coren, S. 1989 'Left-handedness and accident-related injury risk', *American Journal of Public Health*

——, 1992 *The Left-Hander Syndrome: The causes and consequences of left-handedness*, London, John Murray

——, and D. Halpern 1991 'Left-handedness: A marker for decreased survival fitness', *Psychological Bulletin*

——, and ——, 1991 'Hand preference and life span', *New England Journal of Medicine*

——, ed. 1990 *Left-Handedness: Behavioral implications and anomalies*, Amsterdam, North-Holland

——, A. Searleman and C. Porac 1986 'Rate of physical maturation and handedness', *Developmental Neuropsychology*

Coryell, J. F., and G. F. Michel 1978 'How supine postural preferences of infants can contribute toward the development of handedness', *Infant Behaviour and Development*

Coulmas, F. 1989 *The Writing Systems of the World*, Oxford, Basil Blackwell

Crovitz, H. F. 1962 'On direction in drawing a person', *Journal of Consult Psychology*

Dart, R. 1949 'The predatory implemental technique of Australopithecus', *American Journal of Physical Anthropology*

Dawson, J. 1972 'Temne-Arunta hand-eye dominance and cognitive style', *Journal of Psychology*

Deglin, V. L. 1976 'Our split brain', *The UNESCO Courier*, January

DeKay, J. T. 1966 *The Left Handed Book*, New York, Evans

Demarest, J., and L. Demarest 1980 'Does the torque test measure cerebral dominance in adults?' *Perceptual and Motor Skills*

Dennis, W. 1958 'Early Graphic Evidence of Dextrality in Man', *Perceptual and Motor Skills*

Dimond, S. J., and D. A. Blizard 1977 *Evolution and Lateralization of the Brain*, New York Academy of Science

Draaisma, Douwe 2004 'Het einde van de hersenverdoving', *AMC-magazine*, October

Duffy, F. H., G. B. McAnulty and S. C. Schachter 1984 'Brain Electrical Activity Mapping', in Geschwind and Galaburda 1984

Dumont, J. J. 1976 *Leerstoornissen*, vols 1 and 2, Rotterdam, Lemniscaat

Eling, P. 1987 'Waarom is rechts schrijven toch beter dan links?' *Jeugd, School en Wereld*

——, and B. Vreede-Chabot 1989 'Is ons schrift gemaakt voor de rechterhand?' *Tijdschrift voor Orthopedagogiek*

Ellis, S. J., et al. 1998 'Is forced dextrality an explanation for the fall in the prevalence of sinistrality with age? A study in northern England', *Journal of Epidemiological Community Health*

——, P. J. Ellis and E. Marshall 1988 'Hand preference in a normal population', *Cortex*

Engen, A. van 1980 *Leerkrachten en handigheidskeuze. Een inventariserend onderzoek naar de handigheidsproblematiek in klas 1 bij het schrijfonderwijs*, Groningen

——, 1981 *Nota Schrijfdidactiek voor Linkshandigen*, Assen, SABD Drente

——, 1982 *Schrijfdidactiek voor Linkshandigen*, Informatiepunt Basisonderwijs no. 170

Ettlinger, G., and A. Moffett 1964 'Lateral Preferences in the Monkey', *Nature*

Finch, G. 1941 'Chimpanzees' Handedness', *Science*

Fincher, J. 1993 *Lefties*, New York, Barnes and Noble

Flament, F. 1975 *Coordination et prevalence manuelle chez le nourris*son, Aix, Eds CNRS

Flatt, A. E. 2008 'Is being left-handed a handicap? The short and useless answer is "yes and no"' *Baylor University Medical Center Proceedings*

Fritschi, L., M. Divitini et al. 2007 'Left-handedness and risk of breast cancer', *British Journal of Cancer*

Gangestad, S. W., and R. A. Yeo 1997 'Behavioral genetic variation, adaptation and maladaptation: An evolutionary perspective', *Trends in Cognitive Sciences*

Gardner, M. 1982 *The Ambidextrous Universe: Left, Right and the Fall of Parity*, London, Pelican Books

Gaur, A. 1992 *A History of Writing*, New York, Cross River Press

Geraedts, J., et al., eds 2009 *Evolutie zit in je genen, over Darwin en Genomics*, Stichting Bio-Wetenschappen & Maatschappij

Geschwind, N., and A. M. Galaburda 1984 *Cerebral Dominance: The biological foundation*, Harvard University Press

—, and —, 1987 *Cerebral Lateralization: Biological mechanisms, associations and pathology*, Massachusetts, MIT Press

—, and —, 1985 'Cerebral lateralization, biological mechanisms, associations and pathology I: a hypothesis and program for research', *Archives of Neurology*

—, and —, 1985 'Cerebral lateralization, biological mechanisms, associations and pathology II: a hypothesis and program for research', *Archives of Neurology*

—, and —, 1985 'Cerebral lateralization, biological mechanisms, associations and pathology III: a hypothesis and program for research', *Archives of Neurology*

Gesell, A., and L. B. Ames 1947 'The development of handedness', *Journal of Genetical Psychology*

Glick, S. D. 1985 *Cerebral Lateralization in Non-Human Species*, New York, Academic Press

Goldschmidt, T. 2003 *De andere linkerkant: links en rechts in de evolutie*, Amsterdam, Athenaeum – Polak & Van Gennep

Gooch, S. 1980 'Right brain, left brain', *New Scientist*

—, 1980 *The Double Helix of the Mind*, London, Wildwood House

Gordon, C. H. 1987 *Forgotten Scripts: Their ongoing discovery and decipherment. Revised and enlarged edition*, New York, Dorset Press

Gordon, N. 1986 'Left-handedness and learning', *Developmental Medicine and Child Neurology*

Gould, J. 1838–1841 *The Zoology of the Voyage of HMS Beagle Under the Command of Captain Fitzroy, R.N., During the Years 1832 to 1836. Edited and superintended by Charles R. Darwin, Part 3. Birds*, London, Smith, Elder and Co.

Graaf-Tiemersma, M. J. de 1995 *Linkshandigheid en dyslexie: de testosterontheorie voor cerebrale lateralisatie*, diss. Universiteit Utrecht

Groenen, M., and H. J. Megens 2009 'Darwins vinken', in Geraedts et al. 2009

Groff, P., 1964 'Who are the better writers – the left-handed or the right-handed?' *Elementary School Journal*

Gross, K. 1985 *Menschenhand und Gotteshand in Antike und Christentum*, Stuttgart, A. Hiersemann

Haenen-van der Hout, C. G. *Handschrift als signaal: wegwijzers voor ouders voor het vroegtijdig herkennen van problemen in de schrijfontwikkeling*, Deventer, Ankh-Hermes

Hall, J. G. 2003 'Twinning', *The Lancet*

Halpern, D. F., and S. Coren 1988 'Do right-handers live longer?', *Nature*

Harburg, E., A. Feldstein and J. Papsdorf 1978 'Handedness and smoking', *Perceptual and Motor Skills*

Hardyck, C., and L. F. Petrinovitch 1977 'Left-handedness', *Psychological Bulletin*

——, R. Goldman and L. F. Petrinovitch 1975 'Handedness and sex, race and age', *Human Biology*

——, L. F. Petrinovitch and R. D. Goldman 1976 'Left-handedness and Cognitive Deficit', *Cortex*

Harris, J. L. 1990 'Cultural Influences on Handedness: Historical and contemporary theory and evidence', in Coren 1990

Harris, N. 1981 'Left hand, Right hand', *British Medical Journal*

Harvey, T. J. 1988 'Science and Handedness', *British Journal of Educational Psychology*

Hatta, T., and E. Nakatsuka 1976 'A note on the handpreference of the Japanese people', *Perceptual and Motor Skills*

Hecaen, H., and J. de Ajuriaguerra 1964 *Left-Handedness: Manual superiority and cerebral dominance*, New York, Grune and Stratton

——, and J. Sauguet 1971 'Cerebral dominance in left-handed subjects', *Cortex*

Herron, J., ed. 1980 *Neuropsychology of Left-Handedness*, New York, Academic Press

——, D. Galin et al. 1979 'Cerebral specialization, writing posture and motor control of writing in left handers', *Science*

Hewes, G. W. 1949 'Lateral dominance, culture, and writing systems', *Human Biology*

Hicks, R., and M. Kinsbourne 1976 'On the genesis of human handedness: A review', *Journal of Motor Behavior*

Hudson, P. 1982 'Linkshandigheid: een biologische ruk naar rechts', *Psychologie*, September

Huheey, J. 1977 'Concerning the Origin of Handedness in Humans', *Behavior Genetics*

Institoris, H., and J. Sprenger 1487/2005 *Heksenhamer / Malleus maleficarum, vertaald en ingeleid door Ivo Gay*, Voltaire

Jaarsma, P. 1963 *Handen in schilderkunst en praktijk*, Amsterdam, Contact

Jabes, E. 1988 *La memoire et la main*, Montpellier, FataMorgana

Jackson, J. 1905 *Ambidextrality*, London, Kegan Paul

Jackson, J.F.E.I.S. 1905 *Ambidexterity, Or, Two-Handedness and Two-Brainedness*, London, Kegan Paul, Trench, Trübner and Co

Jarman, R. F., and J. G. Nelson 1980 'Torque and cognitive ability: some contradictions to Blau's proposals', *Journal of Clinical Psychology*

Jaynes, J. 1976 *The Origin of Consciousness in the Breakdown of the Bicameral Mind*, Boston, Houghton Mifflin

Jensen, H. 1969 *Die Schrift in Vergangenheit und Gegenwart*, Berlin, VEB deutscher Verlag der Wissenschaften

Johnston, D. W., et al. 2009 'Nature's Experiment? Handedness and early childhood development', *Demography* 46

—, M. Shah and M. A. Shields 2007 *Handedness, Time Use and Early Child Development*, IZA discussion paper 2752, Bonn, IZA

Jordan, H. E. 1914 'Hereditary left-handedness with a note on twinning', *Journal of Genetics*

Kappers, E. J. 1986 *Structureringstendentie, hemisfeerspecialisatie en leren lezen*, diss. Universiteit Utrecht

Kellmer Pringle, M. L. 1961 'The incidence of some supposedly adverse family conditions and of left-handedness in schools for maladjusted children', *British Journal of Educational Psychology*

Kerckhove, D. de, and C. J. Lumsden, eds 1988 *The Alphabet and the Brain: The lateralization of writing*, Berlin, Springer Verlag

Keulen, M. 1981 'Linkshandigen: laten we ze links liggen?' *Jeugd, School en Wereld*

Kimura, D. 1973 'Manual activity during speaking II: Left handers', *Neuropsychologia*

Knecht, S., et al., 2000 'Handedness and hemispheric language dominance in healthy humans', *Brain*

Komai, T., and G. Fukuoka 1934 'A study on the frequency of left-handedness and left-footedness among Japanese schoolchildren', *Human Biology*

Kramer, J. 1961 *Linkshändigkeit: Wesen, Ursachen, Erscheinungsformen, mit Übungen für linkshändige und gehemmte Kinder und Jugendliche*, Solothurn, Antonius Verlag

Laan, M. van der 1991 'Links is toch zo eng niet', *Trouw*, 9 March

Lanthony, P. 2005 *Les peintres gauchers*, Lausanne, Editions L'Âge d'Homme

Lawson, N. C. 1978 'Inverted writing in right- and left-handers in relation to lateralization of face recognition', *Cortex*

Levy, J. 1969 'Possible basis for the evolution of lateral specialization of the human brain', *Nature*

—, 1976 'Review of evidence for a genetic component in the determination of handedness', *Behavior Genetics*

—, and T. Nagylaki 1972 'A model for Genetics of Handedness', *Genetics*

Liederman, J., and M. Kinsbourne 1980 'Rightward bias in neonates depends upon parental right handedness', *Neuropsychologia*

—, and —, 1980 'The mechanism of neonatal rightward turning bias: A sensory or motor asymmetry?' *Infant Behaviour and Development*

—, and J. Coryell 1982 'The origin of left-hand preference: Pathological and non-pathological influences', *Neuropsychologia*

Lishman, W., and E.R.L. McMeekan 1977 'Handedness in relation to direction and degree of cerebral dominance', *Cortex*

Lombroso, C. 1888 *L'homme criminel*, Rome, Bocca frères

—, 'Left-handedness and left-sidedness', *North American Review*

Lovejoy, A. O. 1936 *The Great Chain of Being: A study of the history of an idea*, Harvard University Press

Mandal, M. K., et al., eds 2000 *Side Bias: A neuropsychological perspective*, Dordrecht, Kluwer

McManus, I. C. 1985 *Handedness, Language Dominance and Aphasia: A genetic model*, Cambridge University Press
——, 2002 *Right Hand, Left Hand, The Origins of Asymmetry in Brains, Bodies, Atoms and Cultures*, London, Weidenfeld and Nicolson
——, and M. P. Bryden 1992 'The genetics of handedness, cerebral dominance, and lateralization', in Rapin and Segalowitz 1992
Mengler, W. 2004 'Linkshändigkeit und Streichinstrumentenspiel, eine Annäherung an ein weitgehend unentdecktes Thema', *Das Orchester*
Mesker, P. 1977 *De menselijke hand: een onderzoek naar de ontwikkeling van de handvaardigheid in relatie tot die van de cerebrale organisatie, gedaan bij leesgestoorde kinderen*, Nijmegen, Dekker and Van de Vegt
Miller, E. 1971 'Handedness and the pattern of human ability', *British Journal of Psychology*
Muller, B. 1985 *Linkshandig schrijven? Dominantie-onderzoek voor linker- en rechterhand*, The Hague, BZZTÔH
Mulligan, J., et al. 2001 'Hormones and Handedness', *Hormone Research*
Naitoh, T., and R. Wassersug 1996 'Why are toads right-handed?', *Nature*
——, R. J. Wassersug and R. A. Leslie 1989 'The physiology, morphology, and ontogeny of emetid behavior in anuran amphibians', *Physiological Zoology*
Needham, R., ed., 1973 *Right and Left: Essays on dual symbolic classification*, University of Chicago Press
Neville, A. C. 1976 *Animal Asymmetry*, London, Edward Arnold
Nottebohm, F. 1970 'Ontogeny of bird song', *Science*
Oldfield, R. C. 1969 'Handedness in musicians', *British Journal of Psychology*
——, 1971 'The assessment and analysis of left-handedness: The Edinburgh inventory', *Neuropsychologia*
Parson, B. S. 1924 *Left-handedness: A new interpretation*, New York, MacMillan
Pavert, J. van de 1982 *Links/Rechts: over linksheid, linkshandigheid en linkshandig schrijven*, Amsterdam, Meulenhoff Educatief
Peters, M., and K. Pedersen 1978 'Incidence of left-handers with inverted writing position in a population of 5910 elementary school children', *Neuropsychologia*
Pointer, J. S. 2001 'Sighting dominance, handedness and visual acuity preference: Three mutually exclusive modalities?' *Ophtalmic and Physiological Optics*
Ponte, L. 1988 'What's right about being left-handed?' *Reader's Digest*, September
Porac, C., and I. C. Friesen 2000 'Hand preference side and its relation to hand preference switch history among old and oldest-old adults', *Developmental Neuropsychology*
——, and S. Coren 1981 *Lateral Preferences and Human Behavior*, New York, Springer
Ramadhani, K., S. G. Elias et al. 2005 'Innate left handedness and risk of breast cancer: Case-cohort study', *British Medical Journal*

Ramaley, F. 1913 'Inheritance of left-handedness', *American Naturalist*
Rapin, I., and S. J. Segalowitz, eds 1992 *Handbook of Neurospychology 6: Child Neuropsychology*, Elsevier
Reed, G., and A. Smith 1962 'A further experimental investigation of the relative speeds of left and right-handed writers', *Journal of Genetic Psychology*
Rhoads, J., and A. Darmon 1973 'Some genetic traits in Solomon Island populations', *Journal of Physical Anthropology*
Ride, D. 1940 'Handedness with special reference to twins', *Genetics*
Rousseau, J. J. 1762 *Emile, ou l'education*, Paris
Ruebeck, C. S., J. E. Harrington Jr and R. Moffitt 2006 'Handedness and earnings', *Laterality*
Salk, L. 1973 'The role of the heartbeat in the relations between mother and infant', *Scientific American*
Satz, P. 1972 'Pathological left-handedness: an explanatory model', *Cortex*
——, 'Left-handedness and early brain insult: An explanation', *Neuropsychology*
——, D. Orsini et al. 1985 'The pathological left-handedness syndrome', *Brain and Cognition*
——, K. Achenbach and E. Fennell 1967 'Correlations between assessed manual laterality and predicted speech laterality in a normal population', *Neuropsychology*
Sawyer, C. E., and B. J. Brown 1976 'Laterality and intelligence in relation to reading ability', *Educational Review*
Schachter, S. C., and B. J. Ransil 1996 'Handedness distributions in nine professional groups', *Perceptual and Motor Skills*
Schaik, E. C. van 1984 *De opvoedbaarheid van de rechter hemisfeer: de neuropedagogische en didactische aspecten bij het leren schrijven met de niet-voorkeurshand*, diss. Universiteit van Amsterdam
Shepherd, M. 1989 *The Left-Handed Calligrapher*, Wellingborough, Thorsons
Sitwell, O. 1945 *Left Hand, Right Hand (5 parts)*, London, MacMillan
Sladden, K. 1987 *Left Handed in a Right Handed World: Handedness and associated specific learning difficulties in school and society*, Bath College of Higher Education
Smart, J. L., C. Jeffery and B. Richards 1980 'A retrospective study of the relationship between birth history and handedness at six years', *Early Human Development*
Smits, R. 2000 'Beethoven linksom', NRC *Handelsblad*, 8 January
Sourdive, C. 1984 *La main dans l'Egypte pharaonique: recherches de morfologie structurale sur les objets egyptiens comportant une main*, Bern, Lang
Sperry, R. W. 1979 'The left-handed report', *Which*
Springer, S. P., and G. Deutsch 1993 *Left Brain/Right Brain (4th edn.)*, New York, W. H. Freeman and Co
Steckenfinger, S. A., and A. A. Ghazanfar I 2009 'Monkey visual behavior falls into the uncanny valley', PNAS

Straaten, A. van 1969 *Linkshandigheid*, Leiden, Stafleu

Strien, J. W. van 1988 *Handedness and Hemispheric Laterality*, diss. Vrije Universiteit Amsterdam

——, 1992 'Classificatie van links- en rechtshandige proefpersonen', *Nederlands Tijdschrift voor de Psychologie*

——, 2000 'Genetic, intra-uterine and cultural origins of human handedness' in Mandal et al. 2000

——, 2003 *The Dutch Handedness Questionnaire*, at www.psyweb.nl/homepage/jan_van_strien_files/hquestionnaire_article.pdf

——, et al. 2005 'Increased prevalences of left-handedness and left-eye sighting dominance in individuals with Williams-Beuren syndrome', *Journal of Clinical and Experimental Neuropsychology*

——, A. Bouma and D. J. Bakker 1987 'Birth stress, autoimmune diseases, and handedness', *Journal of Clinical and Experimental Neuropsychology*

Tootal, J. 1990 'Why do humans and apes cradle babies on their left side?' *New Scientist*

Uhrbrock, R. S. 1973 'Laterality in art', *Journal of Aesthetics and Art Criticism*

Vaid, J., U. Bellugi and H. Poizner 1989 'Hand dominance for signing', *Neuropsychologia*

Van Riper, C. 1935 'The quantitative measurement of laterality', *Journal of Experimental Psychology*

Verhaegen, O., and A. Ntumba 1964 'Note on the frequency of left handedness in African children', *Journal of Educational Psychology*

Vermaseren, M. J. 1983 *Corpus Cultus Iovis Sabazii, I: The hands*, Leiden, E. J. Brill

Vlugt, H. van der 1979 *Lateralisatie van hersenfuncties: een neuropsychologisch onderzoek naar de relatie tussen handvoorkeur en de lateralisatie van de taalfunctie*, Lisse, Swets & Zeitlinger

Wagenaar, W. A. 1981 'Als Kuifje naar links beweegt is er iets mis', NRC *Handelsblad*, 5 August

Wassenaar, W. A., and F.P.M. Saan 1983 *De mythe van het linkshandige kind: feiten en ficties in de psychologische diagnostiek*, Nijmegen, Dekker & Van de Vegt

Whitaker, H., and H. A. Whitaker, ed. 1977 *Perspectives in Neurolinguistics* vol. 3, San Diego, Elsevier

Whittington, J. E., and P. N. Richards 1987 'The stability of children's laterality prevalencies and their relationship to measures of performance', *British Journal of Educational Psychology*

Willems, R. M., et al. 2009 'Body-specific motor imagery of hand actions: Neural evidence from right- and left-handers', *Frontiers in Human Neuroscience*

——, M. V. Peelen and P. Hagoort 2009 'Cerebral lateralization of face-selective and body-selective visual areas depends on handedness', *Cerebral Cortex*

Wilson, P. T., and H. E. Jones 1932 'Left-handedness in twins', *Journal of Genetics*

Witelson, S. F. 1985 'The brain connection: The corpus callosum is larger in left-handers', *Science*

Wood, E. K. 1988 'Less sinister statistics from baseball records', *Nature*

Woodson, W. E. 1981 *Human Factors Design Handbook: Information and guidelines for the design of systems, facilities, equipment, and products for human use*, New York, McGraw-Hill

Wrangham, R. 2009 *Koken: over de oorsprong van de mens*, Amsterdam, Nieuw Amsterdam

Wyman, V. 1988 'An ergonomic necessity', *The Engineer*

Young, G., ed., 1983 *Manual Specialization and the Developing Brain*, New York, Academic Press

PHOTO ACKNOWLEDGEMENTS

Photograph ABC Press: p. 129 top; Illustration © Decotype 1989: p. 152; Illustration from Hewes, 'Lateral dominance', 1949: p. 155; Illustration from John Gould, *The Zoology of the Voyage of HMS Beagle*, 1838–41: p. 236; Engraving from Wouter Hutschenruyter, *Grepen uit de geschiedenis van de piano*: p. 91; Illustration from Jensen, *Die Schrift in Vergangenheid und Gegenwart*, 1969: p. 153; Photograph Reuters: p. 129 bottom; Photograph A. Samagalski: p. 122; Illustration from Steckenfinger & Ghazanfar I, 2009: p. 55; US Library of Congress: p. 285.

INDEX